Gesellschaft, Recht, Wirtschaft
Band 3

Gesellschaft, Recht, Wirtschaft

Beiträge aus den Rechts-,
Wirtschafts- und Sozialwissenschaften

Im Auftrag der Universität Mannheim
mit Unterstützung der Stiftung Rheinische Hypothekenbank
herausgegeben von
Eduard Gaugler, Wolfgang Goedecke,
Heinz König, Günther Wiese,
Rudolf Wildenmann

Band 1:
Ökonometrische Modelle und sozialwissenschaftliche
Erkenntnisprogramme
von H. Albert, M. C. Kemp, W. Krelle,
G. Menges, W. Meyer

Band 2:
Essays zur historischen Entwicklung des
Bankensystems
von R. Bogaert und P. C. Hartmann

Band 3:
Problems of Time Series Analysis
by M. Nerlove, S. Heiler, H.-J. Lenz, B. Schips, H. Garbers

Problems of Time Series Analysis

by
Prof. Marc Nerlove Ph.D., Northwestern University,
Evanston, Illinois
Prof. Dr. Siegfried Heiler, Universität Dortmund
Prof. Dr. Hans-J. Lenz, Freie Universität Berlin
Prof. Dr. Bernd Schips, Hochschule
für Wirtschafts- und Sozialwissenschaften St. Gallen
Prof. Dr. Hermann Garbers, Universität Zürich

Bibliographisches Institut Mannheim/Wien/Zürich
B. I.-Wissenschaftsverlag

CIP-Kurztitelaufnahme der Deutschen Bibliothek

Problems of time series analysis / by Marc Nerlove ...
- Mannheim, Wien, Zürich : Bibliographisches
Institut, 1980.
(Gesellschaft, Recht, Wirtschaft; Bd. 3)

NE: Nerlove, Marc [Mitarb.]

Gedruckt mit Unterstützung der Stifung Rheinische Hypothekenbank

ISBN 978-1-4615-9927-2   ISBN 978-1-4615-9925-8 (eBook)
DOI 10.1007/978-1-4615-9925-8

Editors' Introduction

The last decade has witnessed an increased interest in time series analysis. Non-parametric methods like spectral and cross spectral analysis are used to discover regularities in individual time series, relationships between specific components of different time series and leads or lags between those series. Box-Jenkins procedures for the parametric estimation of autoregressive-moving average schemes belong nowadays to the standard equipment of a computer center.

In economics this revival of time series analysis has led to numerous empirical studies on optimal seasonal adjustment procedures, the behavior of prices, production, employment etc. More recently, Box-Jenkins methods form an integral part for tests on the efficiency of markets, the effectiveness of monetary and fiscal policies and for the study of the rôle of different assumptions on the formation of expectations.

This volume comprehends a series of lectures which deal with various topics of time series analysis delivered during the wintersemester 1978/79 at the faculty of economics and statistics. The collection begins with a paper by M. Nerlove introducing the concept of unobserved components. Theoretical results are illustrated by examples for time series on prices of steers, heifers, cows and milk, of cattle and hog slaughter, of industrial production and male unemployment. The study by S. Heiler considers a mixed model with a linear regression part and a regular residual process for the prediction of economic processes when additional information is available. For an autoregressive parametrisation of the residual process theoretical implications are developed in detail. Time series of net production and sales of German clothing and incoming orders as indicator of additional information are used to demonstrate the prediction performance of this approach. H.-J. Lenz discusses the estimation of distributed lags by density methods in case that individual lags can be sampled. In simulation experiments these results are compared with estimates of transfer functions for corresponding aggregate data performed with standard methods. The integration of a so-called observational model in a structural econometric model in order to improve the precisions of short-term forecasts is the subject of the study by B. Schips. The ad-

6

vantage of this approach against an enlargement of the structural model is demonstrated for a small-scale model. Finally, the lecture by H. Garbers discusses principle problems with respect to the search of hypotheses if non-experimental time series data are used for the verification.

Editors have to thank Jürgen Wolters for the preparation of the seminar as well as for his steady effort for the publication of this volume.

Winter 1979   Heinz König

Contents

Contents

Unobserved Components Models for Economic Time Series

by

Marc Nerlove*

Evanston, Illinois, U.S.A.

"It seems very pretty", she said when she had finished it, "but it's *rather* hard to understand! ... Somehow it seems to fill my head with ideas – only I don't exactly know what they are!"

Through the Looking-Glass

* The material in this lecture is drawn largely from my book written with David M. Grether and José L. Carvalho, *Analysis of Economic Time Series: A Synthesis* (New York: Academic Press, 1979). The Figures originally appeared as Figures 1–6, Chapter VI, pp. 110–114; figures 9–16, Chapter XII, pp. 265–269, and Figures 19–24, Chapter XII, pp. 278–289. The table originally appeared as Table 3, Chapter XII, pp. 276–77. The figures and table are reproduced here with the permission of the publisher, Academic Press, Inc., 111 Fifth Ave., New York, N.Y. 10003.

1. Introduction
2. Unobserved-Components Models
3. Estimation and Hypothesis Testing
4. Examples of Estimated Unobserved-Components Models for Eight Series
5. Conclusion
6. References

Zusammenfassung:

Modelle mit nichtbeobachtbaren Komponenten (UC) unterstellen, daß diese durch einfache ARMA-Prozesse repräsentiert werden können, so daß die beobachtete Zeitreihe eine Überlagerung mehrerer ARMA-Prozesse ist. Solche Zerlegungen sind aus der klassischen Zeitreihenanalyse bekannt. Dementsprechend werden in dieser Arbeit einfache ARMA-Modelle für die nichtbeobachtbaren Komponenten „Trend-Zyklus", „Saison" und „irreguläre Einflüsse" für acht ökonomische Zeitreihen behandelt. Als Kriterium für die Modellierung der einzelnen Komponenten dienen die durch die empirische Spektralzerlegung gegebenen Charakteristika dieser Reihen. Die Schätzung der UC-Modelle erfolgt im Frequenzbereich. Dieser Ansatz gestattet es, sehr einfache Modelle mit nicht-beobachtbaren Komponenten für ökonomische Zeitreihen zu formulieren und zu schätzen und diese für Prognosen und für die Aufstellung verteilter lag Modelle zu benutzen. Es ist im Gegensatz zu den ARMA-Modellen für die Gesamtreihe möglich, flexiblere Anpassungen für die einzelnen Komponenten zu erhalten; dies wirkt sich insbesondere für den Saisonteil vorteilhaft aus.

Summary:

An unobserved-components (UC) model for an economic time series simply represents the time series as the superposition of two or more ARMA models, for which the series these represent are not themselves directly observed. A three-component model, with components corresponding to the traditional "trend-cycle", "seasonal" and "irregular", each of which is determined by a relatively simple ARMA process, is capable of reproducing these typical characteristics of an economic time series. In choosing such models for eight economic time series their empirical spectral densities serve as criteria for formulating the UC models. Estimation is done in the frequency domain. The results show that it is possible to construct and estimate simple UC models for economic time series. Such models have many uses, as, e.g. forecasting and formulation of distributed lag models, the most obvious being for seasonal adjustment.

1. Introduction

The idea of decomposing an economic time series into several unobserved components has a long history in economics dating back at least to the early work of Charles Babbage [1] on bank clearings in 1856. The concept of a seasonal component in the demand for money played a major role in the debate preceeding the Bank Reform Act of 1844 in England by which the Bank of England was given the sole right and responsibility to issue bank notes. By the time of W. Stanley Jevons, that economic time series contained certain, perhaps not perfectly regular, periodicities was widely accepted. The methods for discovering and removing periodicities, especially seasonal periodicities which were developed by Jevons [6] and his modern successors, Persons [10], Kuznets [8], Burns and Mitchell [4] and Shiskin [12] among others, have remained essentially the same to these days. Unfortunately, no formal model of the structure of economic time series was ever introduced to justify the operations performed on economic time series, for example, to seasonally adjust them.

Early in this century, the Russian statistician Eugen Slutzky [13] and the British statistician G. U. Yule [14], [15], [16] showed that if we begin with a series of purely random numbers and then take sums of differences, weighted or unweighted, of such numbers, the new series so produced has many of the apparent cyclic properties that were thought at the time to characterize economic and other time series. Slutzky and Yule thus laid the foundation for moving-average processes or autoregressive models for characterizing the serial properties of time series. Such models, or more often combinations of the two, are called ARMA models, or sometimes ARIMA models, and form the basis for the currently popular methods of analyzing and forecasting time series exposited by Box and Jenkins [3][1].

ARMA models may conveniently be represented by means of polynomials in the lag operator U, such that

(1.1) $\qquad U^k x_t = x_{t-k}$,

where x_t, $t = \ldots, -1, 0, 1, 2, \ldots$ is any discrete time series. Let the polynomials in U be defined

(1.2) $\qquad H(U) = \sum_{k=0}^{q} h_k U^k$,

$\qquad\qquad G(U) = \sum_{k=0}^{p} g_k U^k$.

We may consider the complementary polynomials

(1.3) $\qquad H(z) = \sum_{k=0}^{q} h_k z^k$

$\qquad\qquad G(z) = \sum_{k=0}^{p} g_k z^k$

in the general complex variable z. According to the fundamental theorem of algebra $H(z) = 0$ has exactly q roots and $G(z) = 0$ has exactly p. The roots are, in general, complex. The general ARMA model for a time series x_t, $t = \ldots, -1, 0, 1, \ldots$ is

(1.4) $\qquad H(U) x_t = G(U) \varepsilon_t$,

where ε_t, $t = \ldots, -1,0,1, \ldots$ is a time series such that

$$E\varepsilon_t = 0$$

$$E\varepsilon_t \varepsilon_{t'} = \begin{cases} \sigma^2, t = t', \\ 0, t \neq t', \end{cases}$$

which is called *white noise*. Equation (1.4) implies that the time series x_t, $t = \ldots, -1,0,1,2, \ldots$, which we write henceforth as $\{x_t\}$, has zero mean; this can always be assured for any stationary series by removing the sample mean. It is a condition for stationarity that all the roots of $H(z) = 0$ lie *outside* of the unit circle[2]. When this condition holds, a function $1/H(z)$ is well-defined, and we often loosely write

$$(1.5) \qquad x_t = \frac{G(U)}{H(U)} \varepsilon_t$$

so that the ARMA model expresses x_t in terms of a *rational function* in the lag operator U applied to the white noise series $\{\varepsilon_t\}$. The function

$$(1.6) \qquad g(z) = \frac{\sigma^2 G(z) G(z^{-1})}{H(z) H(z^{-1})}$$

is called the autocovariance generating transform for the series $\{x_t\}$[3]. When $g(z)$ is evaluated *on* the unit circle, i.e., at points such as $z = e^{i\lambda}$, it is proportional to the *spectral density function* of the time series $\{x_t\}$[4]. Thus,

$$(1.7) \qquad f(\lambda) = \frac{1}{2\pi} g(e^{i\lambda}) = \frac{\sigma^2 |G(e^{i\lambda})|^2}{2\pi |H(e^{i\lambda})|^2},$$

since $e^{-i\lambda}$ is the complex conjugate of $e^{i\lambda}$. It follows that ARMA models characterize by time series which have *rational spectral density functions,* in the sense that $g(e^{i\lambda})$ is a rational function in $e^{i\lambda}$ and $e^{-i\lambda}$. On the far right hand side of (1.7), the expression there is a special case of what is called the *canonical factorization* of the spectral density function. This factorization is easy to state and use for any ARMA model; we shall indicate how it may be found for any unobserved-components model as defined below. The corresponding ARMA model is then often called the *canonical form* of the unobserved components model.

2. Unobserved - Components Models

An *unobserved - components model* for an economic time series simply represents the time series as the superposition of two or more ARMA models, for which the series these represent are not themselves directly observed. A very simple example taken from Nerlove, *et al.* [9, pp. 73–74] may help to clarify matters. The series $\{x_t\}$ is observed but the series $\{y_t\}$, $\{u_t\}$ and $\{v_t\}$ are not. The model is

$$(2.1) \qquad x_t = y_t + u_t, \qquad y_t = \alpha y_{t-1} + v_t, \qquad |\alpha| < 1,$$

where $\{u_t\}$ and $\{v_t\}$ are independent zero-mean white noise series with variance ratio

$$\mu = Ev_t^2 / Eu_t^2 = \sigma_v^2 / \sigma_u^2.$$

We also assume that $\{u_t\}$ and $\{v_t\}$ for any set of T indices $t_1, \ldots, t_T$ follows a multivariate normal distribution[5]. Using a somewhat loose notation we may rewrite (2.1) as the sum of two simple ARMA models

$$(2.2) \qquad x_t = u_t + \frac{v_t}{1 - \alpha U}.$$

In turn, it can be shown that $\{x_t\}$ also has the representation:

$$(2.3) \qquad (1 - \alpha U) x_t = (1 - \beta U) \varepsilon_t,$$

where $\{\varepsilon_t\}$ is a white noise series with variance $\sigma^2 = (\alpha/\beta)\sigma_u^2$ and where

$$|\beta| = \left| \frac{(1 + \mu + \alpha^2) - [(1 + \mu + \alpha^2)^2 - 4\alpha^2]^{\frac{1}{2}}}{2\alpha} \right| < 1.$$

Consequently, in the canonical form of (2.1) we take

$$G(U) = 1 - \beta U \quad \text{and} \quad H(U) = 1 - \alpha U.$$

The spectral density function for $\{x_t\}$ is

$$(2.4) \qquad f(\lambda) = \frac{\sigma^2}{2\pi} \frac{(1 - \beta e^{i\lambda})(1 - \beta e^{-i\lambda})}{(1 - \alpha e^{i\lambda})(1 - \alpha e^{-i\lambda})}$$

$$= \frac{\sigma^2}{2\pi} \left[\frac{1 + \beta^2 - 2\beta \cos \lambda}{1 + \alpha^2 - 2\alpha \cos \lambda} \right].$$

The reader is invited to graph this function for various values of $|\alpha|$, $|\beta| < 1$ in order to get some feeling for how a spectral density function for so simple a model might look.

A more general model, which allows us to capture much of what Jevons and his modern successors were trying to do with respect to seasonality and other forms of periodicity, is as follows: Let the series $\{x_t\}$ be composed of three parts, which we will call respectively, trend-cycle, seasonal, and irregular.

$$(2.5) \qquad x_t = T_t + S_t + I_t$$

The unobserved components $\{T_t\}$, $\{S_t\}$, and $\{I_t\}$ are assumed to satisfy the relations

$$(2.6) \qquad \begin{cases} Q(U)\,T_t = P(U)\,v_t \\ S(U)\,S_t = R(U)\,w_t \\ \qquad I_t = u_t \,, \end{cases}$$

where $Q(\ )$, $P(\ )$ and $R(\ )$ are polynomials of low order in the lag operator U and $S(\ )$ is a polynomial of low order in the Lth power of U, where L is the number of times per year the series is observed. We assume that $\{v_t\}$, $\{w_t\}$ and $\{u_t\}$ are independent zero-mean with noise series and each follows a normal distribution with variances σ_v^2, σ_w^2 and σ_u^2, respectively. We suppose that a polynomial trend has been removed and that the series $\{x_t\}$ is stationary.

How complicated one may wish to make the polynomials $Q(\ )$, $P(\ )$, $R(\ )$ and $S(\ )$ depends in part on how many data points are available and on how closely one wishes to approximate the characteristics of the original series. The theoretical spectral density of the model (2.6) is

$$(2.7) \qquad f(\lambda) = \frac{1}{2\pi} \left\{ \frac{|P(e^{i\lambda})|^2}{|Q(e^{i\lambda})|^2} \sigma_v^2 + \frac{|R(e^{i\lambda})|^2}{|S(e^{i\lambda})|^2} \sigma_w^2 + \sigma_u^2 \right\} .$$

Clearly the fraction of the total variance of the series $\{x_t\}$ contributed by each component depends not only on the variance σ_v^2, σ_w^2, and σ_u^2 but also on the parameters of the polynomials $P(\ )$, $Q(\ )$, $R(\ )$ and $S(\ )$.

Figure 1 graphs two rather typical spectral densities, U.S. automotive sales and inventories. The estimates exhibit two features commonly found: First, there are peaks at each of the so-called seasonal frequencies (with monthly data these are frequencies corresponding to 1, 2, 3, 4, 5, and 6 cycles per year). Second, apart from these seasonal peaks, the spectral densities are generally decreasing with frequency, having substantially more power at the frequencies near zero than anywhere else.

A three-component model, with components corresponding to the traditional "trend-cycle", "seasonal", and "irregular", each of which is represented by a relatively simple mixed moving-average, autoregressive scheme, is capable of reproducing these typical characteristics of an economic time series. In choosing such a model we seek one which is simple, i. e., has only a few parameters, and one which has a spectral density with characteristics which can be associated with particular unobserved components of the traditional kind.

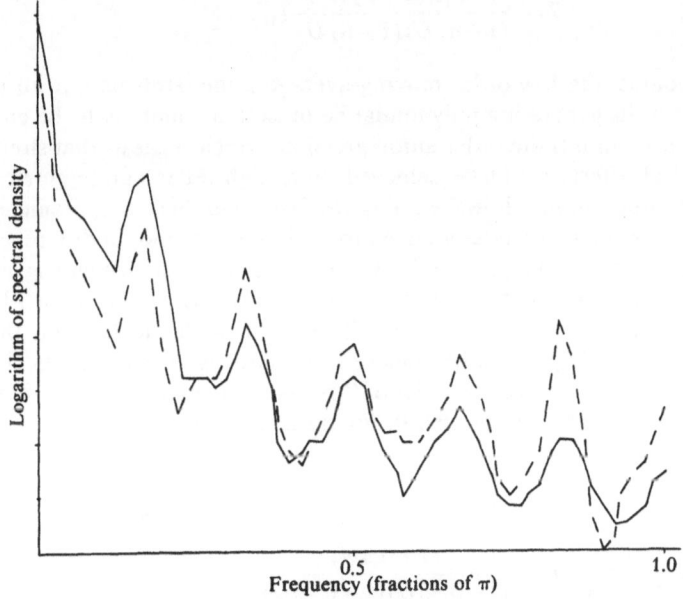

Figure 1: Spectral densities: - - - -, automotive sales; ———, automotive inventories.

Thus, the "trend-cycle" should be a series, the time pattern of which is dominated by gradual cumulative movements and does not have any prominent short-term regularities. In terms of its spectral density, it should have maximum power at the origin and decreasing power throughout its range. Many low-order autoregressive processes have spectra of the type described. There are two points worth mentioning, however, in connection with economic time series: First, in undifferenced form very little of the variance appears to be contributed by what we have called the trend-cycle component at high frequencies.

Much more of a contribution is made at those frequencies by the seasonal and irregular. Purely autoregressive processes of low order, however, with real roots have spectral densities which decline less slowly with frequency than is desirable if we wish our model to allow greater contribution of seasonal and irregular components at high frequencies. In order to produce this effect it is necessary to formulate the trend-cycle component as a mixed process with a low-order moving-average as numerator and a low-order autoregression as denominator, for example:

$$(2.8) \qquad T_t = \frac{(1 + \beta_1 U + \beta_2 U^2)}{(1 - \alpha_1 U)(1 - \alpha_2 U)} \varepsilon_{1t}.$$

Of course, the low-order moving-average numerator may, if all the roots of its generating polynomial lie outside the unit circle, be equivalent to an infinite order autoregression, which suggests that similar spectral effects could be achieved with high-order autoregressions. High-order terms, however, require large numbers of parameters; thus, parsimony dictates a mixed formulation. Second, in differenced form, many of our price series have a spectral shape which suggests stationarity in this form (absence of a sharp dip at the origin) but nonetheless have a spectral density with a moderate fall and then arise. Such shapes can be achieved by allowing complex roots. The graphs of the spectral densities of three possible models for the trend-cycle component are plotted in Figure 2. They are

$$T_t = \frac{1}{(1 - 0.95 U)(1 - 0.75 U)} \varepsilon_{1t}$$

$$(2.9) \qquad T_t = \frac{(1 + 0.8 U)}{(1 - 0.95 U)(1 - 0.75 U)} \varepsilon_{1t}$$

$$T_t = \frac{1}{(1 - 1.7 U + 0.9025 U^2)} \varepsilon_{1t},$$

where $\{\varepsilon_{1t}\}$ is white noise.

Note that the last possibility has complex roots, namely $0.9418 \pm i \sqrt{0.2210}$, which lie outside the unit circle.

The simplest process with peaks at each of the seasonal frequencies is

$$(2.10) \qquad S_t = \frac{1}{1 - \gamma U^L} \varepsilon_{2t}, \qquad \text{where } \{\varepsilon_{2t}\} \text{ is white noise, and}$$

where L is the number of observations per year. A process of this type amounts to specifying L independent processes each of which appears as a first-order autoregression when observed at annual intervals. Depending on whether γ is close to one or close to zero, the seasonal peaks produced will be either extremely sharp or very diffuse.

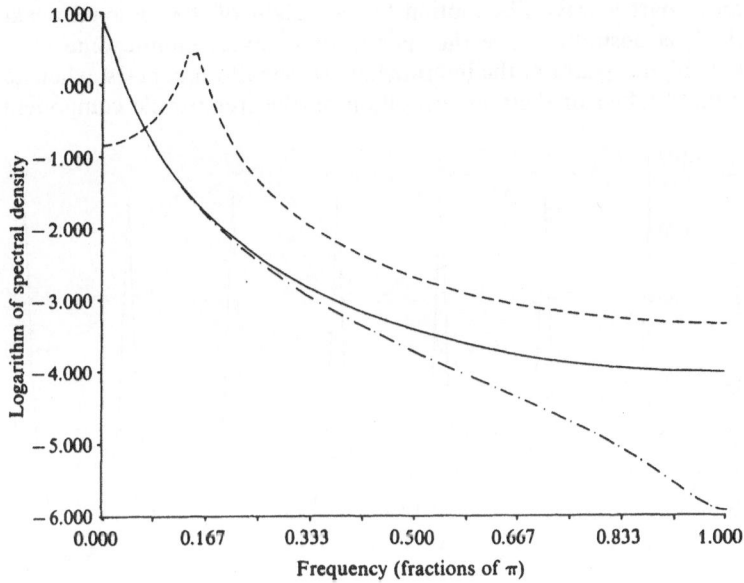

Figure 2: Spectral densities: ———, $T_t = \dfrac{\varepsilon_{1t}}{(1-0.95U)(1-0.75U)}$,

————, $T_t = \dfrac{(1+0.8U)\varepsilon_{1t}}{(1-0.95U)(1-0.75U)}$,

- - - -, $T_t = \dfrac{\varepsilon_{1t}}{(1-1.7U+0.9025U^2)}$.

Figure 3 shows the spectral density of S_t ($\gamma = 0.9$, $L = 12$); note that the peaks at each seasonal frequency are the same height, which is not characteristic of real economic time series. In undifferenced form most economic time series exhibit spectral densities with markedly higher peaks at the lower seasonal frequencies than at the higher. Merely superimposing a seasonal component with peaks of equal height on the component T_t as defined in (7) will not produce the effect characteristic of many economic time series; it is necessary, as be-

fore, to introduce a moving-average numerator in order to achieve
this effect. Thus,

$$(2.11) \qquad S_t = \frac{1 + \beta_3 U + \beta_4 U^2}{1 - \gamma U^{12}} \varepsilon_{2t}$$

should be used. A second-order moving-average in the numerator
can impart a wave-like motion to the height of the seasonal peaks,
which is desirable since the spectra of many economic time series
show higher peaks at the intermediate seasonal frequencies, when account is taken of their superposition on the trend-cycle component,

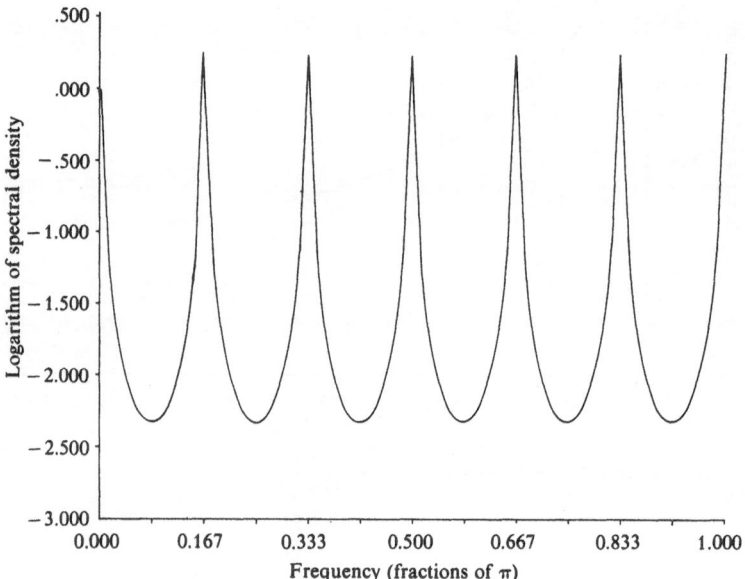

Figure 3: Spectral density: ———, $S_t = \varepsilon_{2t}/(1 - 0.9U^{12})$.

than at either very high or at very low frequencies. A graph of the
theoretical spectral densities series with seasonal component

$$(2.12) \qquad S_t = \frac{(1 + 0.6U)}{(1 - 0.8U^{12})} \varepsilon_{2t}, \quad \text{and}$$

$$S_t = \frac{(1 + 0.8U)(1 - 0.8U)}{(1 - 0.8U^{12})} \varepsilon_{2t},$$

are presented in Figure 4.

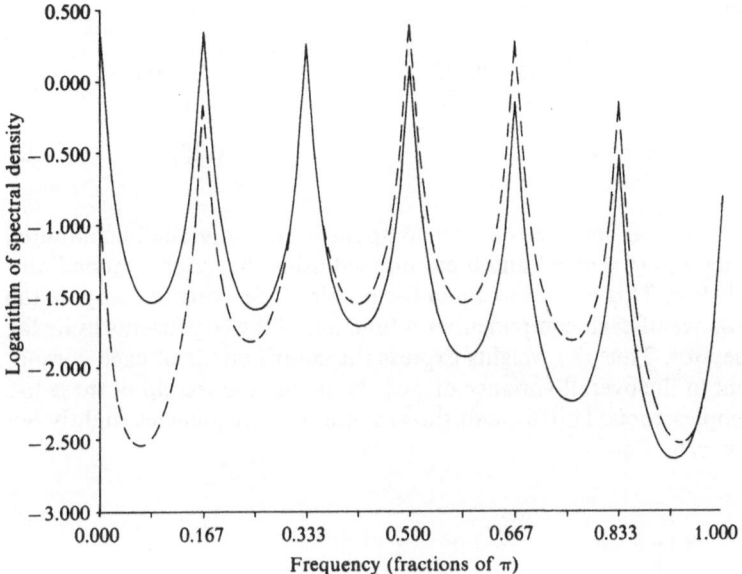

Figure 4: Spectral densities: ———, $S_t = \dfrac{(1 + 0.6\,U)\,\varepsilon_{2t}}{(1 - 0.8\,U^{12})}$,

————, $S_t = \dfrac{(1 + 0.8\,U)(1 - 0.8\,U)\,\varepsilon_{2t}}{(1 - 0.8\,U^{12})}$.

These considerations lead to the view that a model such as

$$(2.13) \qquad x_t = \frac{(1 + \beta_1\,U + \beta_2\,U^2)}{(1 - \alpha_1\,U)(1 - \alpha_2\,U)}\,\varepsilon_{1t}$$
$$+ \frac{(1 + \beta_3\,U + \beta_4\,U^2)}{(1 - \gamma U^{12})}\,\varepsilon_{2t} + \varepsilon_{3t}$$

may be suitable for the representation of many economic time series. The graph of the theoretical spectral densities for two such models are presented in Figures 5 and 6. The models are

$$(2.14) \qquad x_t = \frac{(1 + 0.8\,U)}{(1 - 0.95\,U)(1 - 0.75\,U)}\,\varepsilon_{1t} + \frac{(1 + 0.6\,U)}{(1 - 0.9\,U^{12})}\,\varepsilon_{2t} + \varepsilon_{3t}$$

weight = (0.85) (0.10) (0.05)

and

$$x_t = \frac{(1+0.6U)}{(1-1.6U+0.8U^2)}\varepsilon_{1t} + \frac{(1+0.6U)(1-0.2U)}{(1-0.8U^{12})}\varepsilon_{2t} + \varepsilon_{3t}$$

weight =   (0.70)   (0.20)   (0.10).

The spectral densities of each component are individually computed using $\sigma_1^2 = \sigma_2^2 = \sigma_3^2 = 1$ and then summed using the weights immediately below. This is done to avoid the problem of having to compute the variance of each component as a function of the coefficients in its lag operator. Thus the weights express the contributions of each component to the overall variance of $\{x_t\}$. Note that the second of these has complex roots, $1 \pm 0.5\,i$, and shows a hump at frequencies slightly below $\pi/6$.

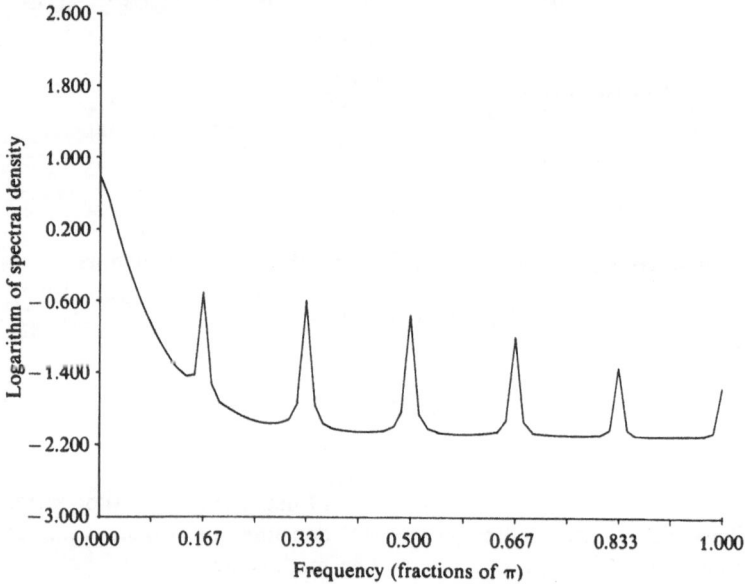

Figure 5: Spectral density:

$$x_t = \frac{(1+0.8U)\,\varepsilon_{1t}}{(1-0.95U)(1-0.75U)} + \frac{(1+0.6U)\,\varepsilon_{2t}}{(1-0.9U^{12})} + \varepsilon_{3t}\,.$$

(weight 0.85)   (weight 0.10)   (weight 0.05)

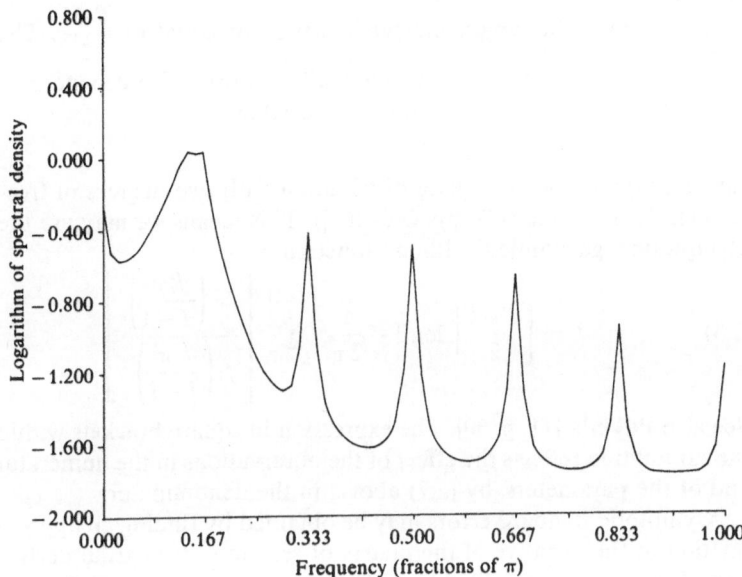

Figure 6: Spectral density:

$$x_t = \frac{(1+0.6U)\,\varepsilon_{1t}}{(1-1.6U+0.8U^2)} + \frac{(1+0.6U)(1-0.2U)\,\varepsilon_{2t}}{(1-0.8U^{12})} + \quad \varepsilon_{3t}\,.$$

(weight 0.70)   (weight 0.20)   (weight 0.10)

3. Estimation and Hypothesis Testing

To estimate an unobserved-components model we make use of the distributional properties of the periodogram ordinates for the observed series $\{x_t\}$. The periodogram ordinates are defined as

$$(3.1) \qquad I_T(\lambda) = \frac{2}{T} \left| \sum_{t=1}^{T} e^{i\lambda t} x_t \right|^2$$

where T is the total number of observations on $\{x_t\}$ in the sample and where

$$(3.2) \qquad \lambda = \frac{k\pi}{T-1}, \quad k=0, 1, \ldots, \left[\frac{T-1}{2}\right].$$

$\left[\dfrac{T-1}{2}\right]$ means the largest integer less than or equal to $\dfrac{T-1}{2}$. The periodogram ordinates are asymptotically distributed independently (Brillinger [2, p. 95]), and the random variables

$$2 I_T(\lambda)/2\pi f(\lambda)$$

have an asymptotic chi-square distribution with two degrees of freedom (L. H. Koopmans [7, pp. 260–265]). This means we may use the asymptotic logarithmic likelihood function

$$(3.3) \qquad L = \left[\frac{T+1}{2}\right]\log 2 - \frac{1}{2\pi}\sum_{k=0}^{\left[\frac{T-1}{2}\right]}\left[\frac{I_T\left(\dfrac{k\pi}{T-1}\right)}{f\left(\dfrac{k\pi}{T-1}\right)}\right].$$

See also Pukkila [11, p. 50]. The expression in square brackets within the summation reflects the effect of the observations in the numerator and of the parameters, by (2.7) above, in the denominator.

Asymptotic standard errors may be obtained by finding an approximation to the negative of the matrix of second-order partial derivative of the logarithmic likelihood function and inverting it. Tests of more complex hypotheses may be made by the usual likelihood ratio test statistics.

The function L is a nonlinear function of the parameters, e.g., those in (2.13), so that L will, in general, have to be maximized by numerical methods. For details the reader is refered to Nerlove, *et al.* [9, pp. 132–139 and 270–274].

4. Examples of Estimated Unobserved-Components Models for Eight Series

A model of the form (2.13) was estimated for eight detrended series as follows:

Series	Period	Trend Removed
Price of steers	1910–74	linear
Price of cows	1925–74	quadratic
Price of heifers	1935–74	quadratic
Price of milk	1910–74	quadratic
Cattle slaughter	1947–74	quadratic
Hog slaughter	1910–74	linear
Industrial production	1919–74	quadratic
Male unemployment, 20+	1947–74	cubic

Figures 7–14 present the spectral densities for the eight detrended series estimated at 72 points in addition to the origin.

The estimates of the densities at each frequency are connected by a solid line. The points above and below each estimated density represent points one standard error of estimate away. The numerically marked points along the abscissa are the seasonal frequencies as fractions of π. It may be noted that all of the spectral densities have the same general shape and form.

All series show some evidence of seasonality, although this is most marked for Cattle and Hog Slaughter, Industrial Production and Male Unemployment, 20+. Curiously, the price series show much less evidence of seasonality except for the rather marked peak at 0.167 which corresponds to a sinusoidal fluctuation with a period of 12 months, in the Milk price and Cow price series. The absence of much evidence of seasonality in the price of a storable commodity would not be at all surprising, but meat and milk were not easily storable throughout much of the period under consideration. Moreover, the peak at 12 months cannot be explained in this manner. Apart from this somewhat irregular evidence of seasonality, each series exhibits a spectral density which falls off rapidly with frequency, suggesting a trend-cycle component with a first- or second-order autoregression tempered by a moving-average. There is little or no suggestion in any of the spectral densities that an autoregressive component in the trend-cycle with complex roots may be found, but we nonetheless allow for such a possibility in estimation, and, indeed, find such complex roots in a number of cases.

Table 1 summarizes the results of fitting the common model (2.13) to all series by the frequency domain methods described in section 3.

For each series the coefficients of all moving-average and autoregressive parts are presented, together with estimates of variances, and the percent contribution of each component to the variance of the observed series is given immediately above the estimated variance of the innovation of that component (*not* equal to the variance of the component). Since complex roots were allowed in the autoregressive part of the trend-cycle component, this was estimated in the form

$$1 + a_1 U + a_2 U^2$$

rather than in the form

$$(1 - \alpha_1 U)(1 - \alpha_2 U).$$

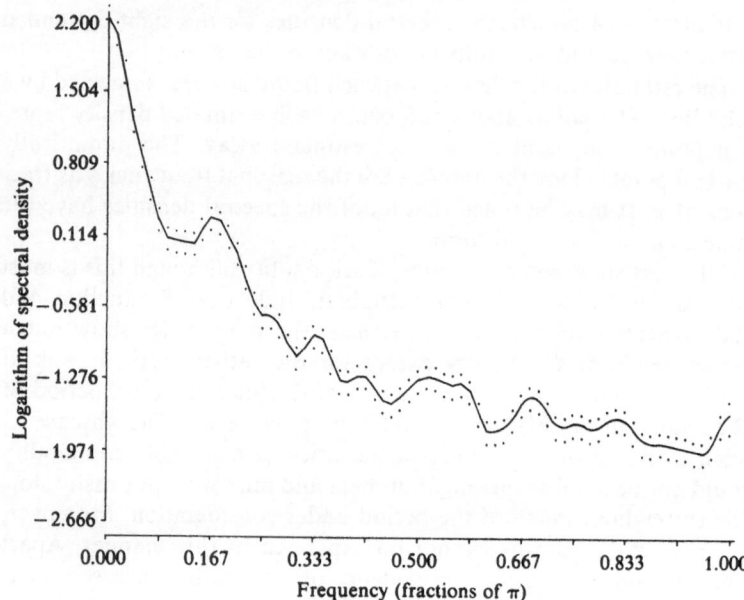

Figure 7: Estimated spectral density, detrended price of steers, 1910–1974.

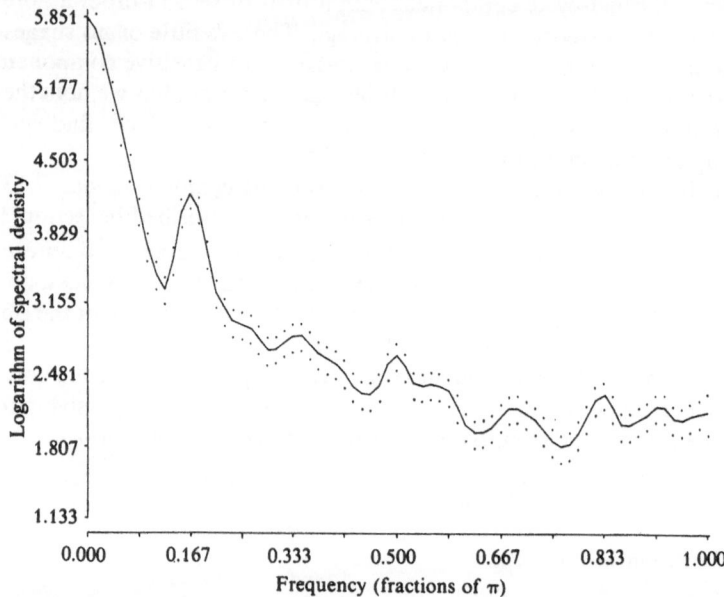

Figure 8: Estimated spectral density, detrended price of cows, 1925–1974.

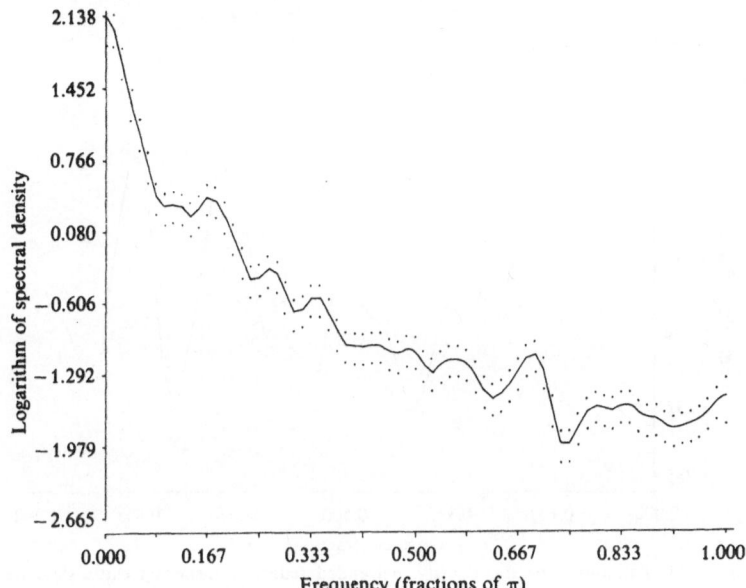

Figure 9: Estimated spectral density, detrended price of heifers, 1935–1974.

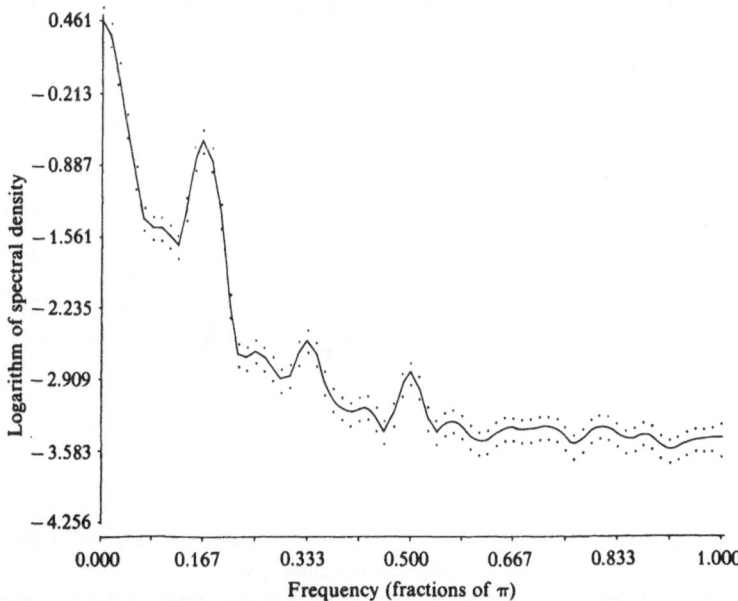

Figure 10: Estimated spectral density, detrended price of milk, 1910–1974.

M. Nerlove

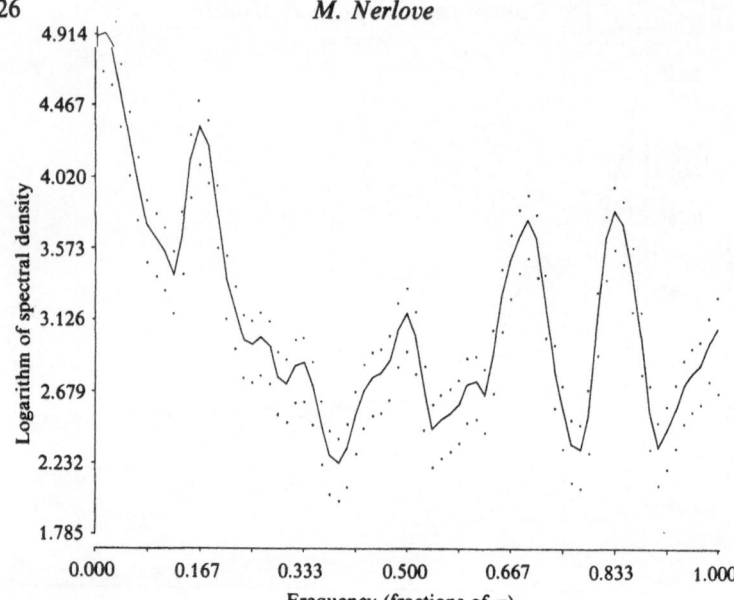

Figure 11: Estimated spectral density, detrended federally inspected cattle slaughter 1947–1974.

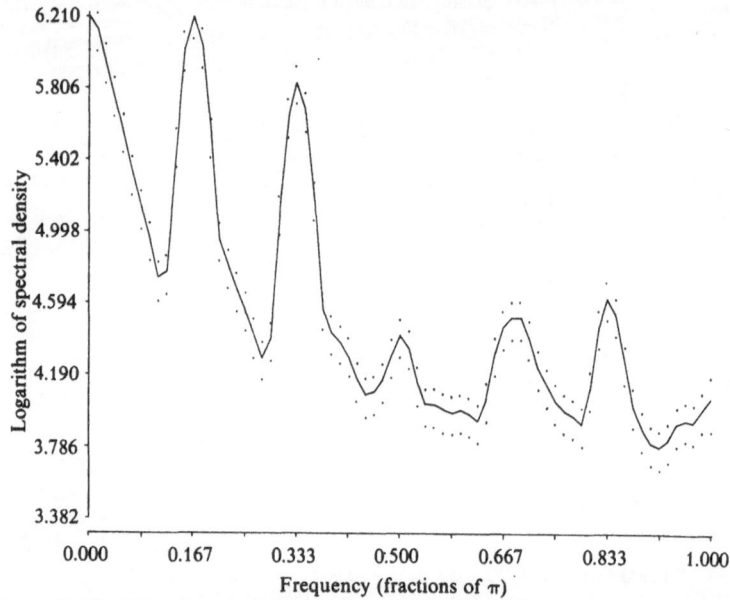

Figure 12: Estimated spectral density, detrended federally inspected hog slaughter, 1910–1974.

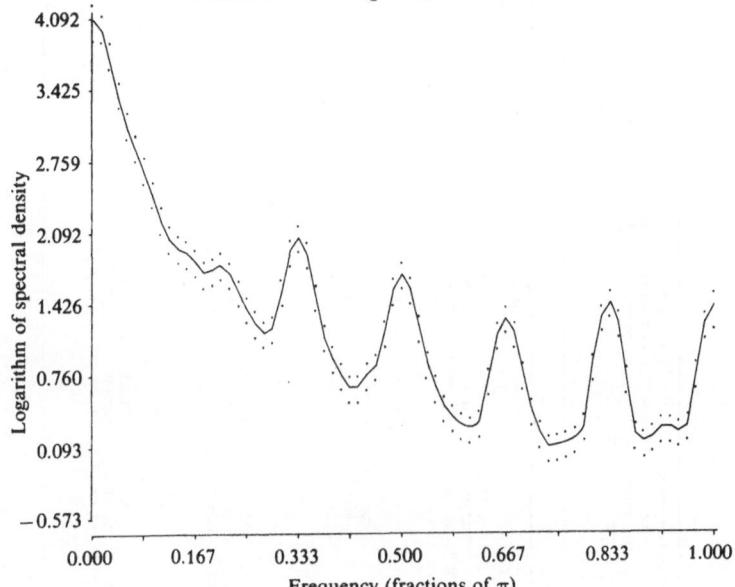

Figure 13: Estimated spectral density, detrended industrial production, 1919-1974.

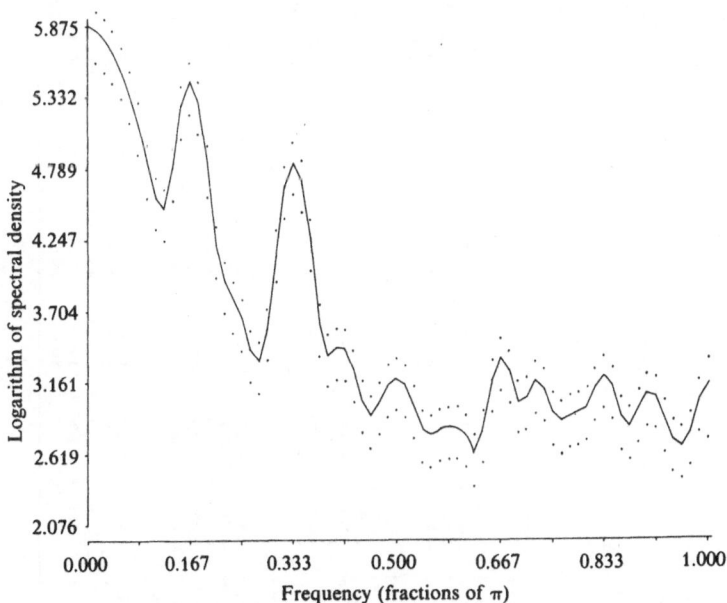

Figure 14: Estimated spectral density, detrended male unemployment, 20+, 1947-1974.

Table 1: Frequency-Domain Estimates of Unobserved-Components Models for Eight-Detrended Series

(1) Series, Period of Fit Item	Trend Cycle Component					Seasonal Component				Irregular Component Variance	Value of the Likelihood Function at the Maximum	Number of Iterations to Maximum
	Moving-Average		Autoregression		Variance	Moving Average		Auto-regression	Variance			
	β_1	β_2	α_1	α_2	σ^2	β_3	β_4	$-\gamma$	σ^2			
	(2)	(3)	(4)	(5)	(6)	(7)	(8)	(9)	(10)	(11)	(12)	(13)
1. Price of Steers 1910–64					99.5%				0.5%	0.0%		
Coefficients	0.8068	0.9943	−1.1841	0.2117	0.1369	−.02676	−.4158	−.1039	0.1069	0.000934	−182.4015	29
Standard Error	1.6075	2.5246	0.1832	0.1781	0.3784	2.2791	1.4254	0.1318	0.3134	0.1568		
Gradient	7.3×10^{-5}	-1.0×10^{-4}	5.6×10^{-4}	4.6×10^{-4}	—	1.2×10^{-4}	-2.7×10^{-4}	-6.6×10^{-5}	3.7×10^{-6}	-2.3×10^{-5}		
Roots	−.4057	± 0.9171i	4.5579	1.0366		−1.5833	1.5189					
Modulus	1.0029		Same			1.5833	Same					
2. Price of Cows[a] 1925–64					97.9%				1.6%	0.5%		
Coefficients	0.1309	−.7583	−1.8934	0.8984	737.912	1.1131	0.99998	−.9579	46.553	628.023	−2121.2394	20
Standard Error	55.3166	44.6897	0.0442	0.0433	42097.	0.2390	0.2853	0.0181	23.712	1175.76		
Gradient	-5.7×10^{-4}	-5.2×10^{-4}	4.7×10^{-2}	4.7×10^{-2}	—	1.9×10^{-4}	-3.4×10^{-4}	-5.2×10^{-4}	-6.2×10^{-4}	2.1×10^{-4}		
Roots	−1.0653	1.2379	1.0538	± 0.0516i		−.5565	± 0.8308i					
Modulus	1.0653	Same		1.0550			1.000008					
3. Price of Heifers 1935–64					98.6%				1.0%	0.4%		
Coefficients	0.7981	−.1579	−1.2271	0.2565	0.3766	1.1978	0.7467	−.9161	0.01245	0.1031	−173.9255	28[b]
Standard Error	98.602	16.284		0.3900	37.806	0.5590	0.6205	0.0581	0.01231	0.6285		
Gradient	-9.3×10^{-6}	1.1×10^{-4}	7.2×10^{-4}	8.4×10^{-4}	—	6.2×10^{-5}	1.2×10^{-5}	-2.7×10^{-4}	-4.9×10^{-4}	1.5×10^{-4}		
Roots	−1.0392	6.0929	3.7417	1.0419		−.8020	± 0.8342i					
Modulus	1.0392	Same		Same			1.1572					
4. Price of Milk 1910–64					94.1%				5.4%	0.5%		
Coefficients	0.6491	0.8801	−1.4851	0.4965	0.000784	1.5163	1.0000	−.9687	0.000355	0.0028	1191.7266	30[c]
Standard Error	19.530	41.112	0.0983	0.0972	0.0376	0.2373	0.1945	0.0106	0.000114	0.00173		
Gradient	-2.3×10^{-4}	-2.2×10^{-4}	4.8×10^{-4}	-4.4×10^{-4}	—	-9.1×10^{-5}	1.3×10^{-5}	4.2×10^{-3}	4.2×10^{-4}	3.6×10^{-5}		
Roots	−.3688	± 1.0001i	1.9672	1.0238		−.7581	± 0.6521i					
Modulus	1.0559		Same				1.00005					

Table 1: (cont.) Frequency-Domain Estimates of Unobserved-Components Models for Eight-Detrended Series

Series, Period of Fit / Item	Trend Cycle Component Moving-Average β_1	β_2	Autoregression α_1	α_2	Variance σ^2	Moving Average β_3	β_4	Seasonal Component Autoregression $-\gamma$	Variance σ^2_r	Irregular Component Variance	Value of the Likelihood Function at the Maximum	Number of Iterations to Maximum
(1)	(2)	(3)	(4)	(5)	(6)	(7)	(8)	(9)	(10)	(11)	(12)	(13)
5. Cattle Slaughter 1947-64					63.1%				36.9%		-1091.6112	16^d
Coefficients	0.7012	$-.2988$	$-.3677$	$-.5594$		$-.5077$	1.0000	$-.8692$		Suppressed		
Standard Error	0.3147	0.1555	0.2222	0.2103	2735.9	0.1749	0.4717	0.0447	1565.0	No convergence for		
Gradient	-1.3×10^{-6}	-2.8×10^{-6}	2.8×10^{-5}	2.8×10^{-5}	1153.9	$-5.2\times10^{-7} \times 4.0\times10^{-7}$		9.2×10^{-7}	837.6	positive		
Roots	-1.00005	3.3471	1.7054	1.7054		$0.2538 + 0.9672i$		8.6×10^{-7}		value when		
Modulus	1.00005	Same	1.7054	Same		1.000002				included.		
6. Hog Slaughter 1919-64					45.4%				50.9%	3.7%	-4336.3101	22
Coefficients	0.7253	$-.2699$	0.2366	$-.6477$		1.0882	0.7533	$-.9656$		34938.		
Standard Error	148.55	37.726	0.1904	0.1720	73506.8	0.1923	0.1885	±0.01124	11915.0	384050.		
Gradient	-1.8×10^{-4}	-3.8×10^{-4}	8.1×10^{-4}	5.2×10^{-4}	1119696.	$1.6\times10^{-7} -3.3\times10^{-10}$	-3.3×10^{-10}	1.3×10^{-3}	3492.9	-3.2×10^{-4}		
Roots	-1.00383	3.6914	-1.4385	1.0733		$-.7223 + 0.8997i$	0.8997i		-3.2×10^{-4}	2.9×10^{-4}		
Modulus	1.00383	Same	1.4385	Same			1.1522					
7. Industrial Production 1919-64					95.5%				4.2%	0.3%	-1502.0965	47
Coefficients	0.4490	$-.5166$	-1.6764	0.6887		0.5949	0.2810	$-.9280$		6.7140		
Standard Error	68.998	36.492	0.2537	0.2466	20442	0.1431	0.1457	0.0215	10.0998	13.2997		
Gradient	-5.0×10^{-4}	-7.7×10^{-4}	6.1×10^{-2}	6.1×10^{-2}	14259	4.8×10^{-6}	3.0×10^{-5}	2.5×10^{-5}	2.3746	1.4×10^{-4}		
Roots	-1.0230	1.8922	1.3883	1.0459		$-1.0587 + 1.5616i$	1.5616i		9.7×10^{-5}			
Modulus	1.0230	Same	Same	Same			1.8866					
8. Male Unemployment 1947-64					80.8%				17.8%	1.3%	-1200.7316	46
Coefficients	0.7799	$-.1162$	-1.7028	0.7308		1.4145	1.0000	$-.9138$		4384.3		
Standard Error	158.142	21.5693	0.2388	0.2258	1460.3	0.4773	0.6546	0.0296	2449.4	4916.9		
Gradient	-2.9×10^{-5}	-3.1×10^{-5}	2.4×10^{-3}	2.3×10^{-3}	233854.	2.3×10^{-5}	2.2×10^{-5}	1.9×10^{-5}	1773.4	5.6×10^{-6}		
Roots	-1.1014	7.8149	$1.1650 + -.1056i$			$-.7074 + 0.7068i$	0.7068i		4.1×10^{-6}			
Modulus	1.1014	Same	1.1697				1.00004					

a Transformed to $(1-0.1U^{12})x_t$, Estimation did not converge for untransformed detrended series x_t.
b Converged in 18 iterations with a non-negative definite Hessian and a non-invertible moving-average in the trend-cycle. The root was inverted and calculations restarted. Convergence was obtained in 10 more iterations with a negative definite Hessian. The value of the likelihood function was exactly the same as before.
c Converged in 21 iterations with a non-negative definite Hessian and a non-invertible moving-average in the trend-cycle. The root was inverted and calculations restarted. Convergence was obtained in 9 more iterations with a negative definite Hessian. The value of the likelihood function was exactly the same as before.
d After restart from last values for which $\hat{\sigma}^2_I > 0$.

The roots, whether real or complex, are presented in the next to last line of the block for each series and the modulus of the root follows in the next line. In the line below the estimated coefficients, estimates of the corresponding standard errors are given. These estimates are calculated from an analytically derived matrix of second-derivatives of the logarithmic likelihood function. Following, the estimated standard errors of the final values of the gradients are given. The value of the likelihood function at the maximum is given in column 12 and the number of iterations to convergence in column 13. Convergence was presumed to have occurred *either* when the largest gradient value became very small *or* when the step-size fell below a preassigned limit. Since the step-size depends in part on how much the value of the function being minimized changes between iterations, and since the likelihood function was frequently flat, calculations were terminated by the step-size condition.

In only one case, Cattle slaughter, were we unable to estimate a three-component model. Moreover, in general the gradients are very small, both relative to the value of the logarithmic likelihood function at the maximum and relative to the size of the coefficients themselves. Unfortunately, with few exceptions, the estimated standard errors of the coefficients tend to be rather large, suggesting that the likelihood function is quite flat in the vicinity of the maximum. Such a conclusion may also be drawn from the fact that, even where different to a notable degree, models estimated in the time domain forecast about as well as models estimated in the frequency domain. Apparently, considerable variation in the parameters is required before either forecasting ability or the value of the likelihood of the sample can be much affected.

The *Price of Steers,* 1910–64, shows a strong and reliable estimated first-order autoregression in its trend-cycle component, which partly as a consequence, contributes 99.5% of the variance of the observed series.

The second-order moving-average in the trend-cycle component is not significantly different from 1. A mild seasonal is apparent from the estimate of the autoregressive parameter in the corresponding component of about 0.10, this value is not large enough to produce a pronounced seasonal effect and its effect is further reduced by the small estimated variance of this component. Here, we were successful in estimating an irregular component ($\sigma_3^2 > 0$), but because of the larger variances attached to the other components, and especially because of the strong autoregression in the trend-cycle, the contribution of this component was effectively zero. A graph of the theoretical spectral density for the estimated model is presented in Figure 15. Disregarding the scale which is essentially arbitrary for the theoretical spectral density, the model provides only a mediocre fit: The first order autoregression is far too strong, producing a rapid fall off in the spectral density for the model which does not occur for the series itself. The mild peak in the actual series at $\pi/6$ is also almost totally suppressed in the model. Correspondingly, the model has somewhat excessive power at frequencies between $\pi/3$ and $2\pi/3$.

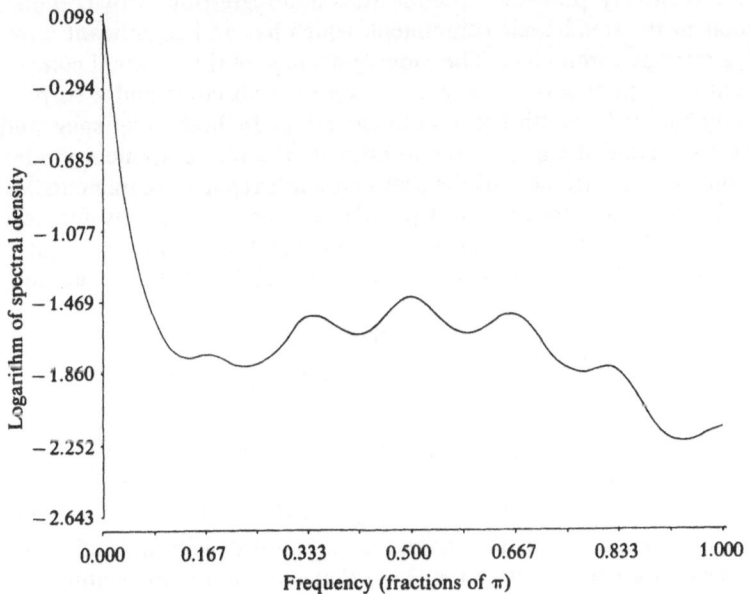

Figure 15: Theoretical spectral density, model for detrended price of steers estimated in the frequency domain, 1910–1964.

The *Price of Cows*, 1925–64, offered the greatest difficulty in estimation: Convergence simply could not be obtained with a positive variance for the seasonal component; moreover, when σ_2^2 was set to zero, σ_3^2 also went to zero, and the resulting ARMA model had estimated innovations containing a strong and significant seasonal component. As may be seen from Figure 8, the series itself has relatively little seasonality except for a rather marked peak at $\pi/6$. The partial autocovariance at lag 12 in the estimated innovations of a simple ARMA model of the form

$$x_t = \frac{1 + \beta_1 U + \beta_2 U^2}{(1 - \alpha_1 U)(1 - \alpha_2 U)} \varepsilon_t$$

fitted to the detrended series was about 0.10. Suppose then we assume a common autoregression in all components of the form $1/(1 - 0.1 U^{12})$; this suggests transforming the data to $(1 - 0.1 U^{12}) x_t$. When this was done the fitted models presented in Table 1 were obtained. The time and frequency domain estimates differ little. There is a relatively powerful second-order autoregression with complex roots in the trend-cycle component, which has an insignificant moving-average component. The moving-average of the seasonal component has complex roots very nearly on the unit circle and a surprisingly marked twelfth-order autoregression. In both frequency and time domains, it was possible to estimate a three-component model, although the variances of the seasonal and irregular components are quite small, so that these components account respectively for only 1.6% and 0.5% of the variance in the model estimated in the frequency domain. A graph of the theoretical spectral density of the theoretical model

$$x_t = \frac{1 + 0.1209 U - 0.7583 U^2}{(1 - 1.8934 U + 0.8984 U^2)(1 - 0.1 U^{12})} \varepsilon_{1t}$$
$$+ \frac{1 + 1.1131 U + 0.9998 U^2}{(1 - 0.9579 U^{12})(1 - 0.1 U^{12})} \varepsilon_{2t} + \frac{1}{(1 - 0.1 U^{12})} \varepsilon_{3t}$$

is given in Figure 16. The fit at the upper end of the spectrum of frequencies is good, but the theoretical spectral density falls off much too rapidly between 0 and $\pi/3$, so that, despite an unusually sharp peak at $\pi/6$, the theoretical density badly misses the marked peak at $\pi/6$ in the estimated density of the actual series. The problem again appears to lie with an overly powerful autoregressive part in the trend-cycle component.

The *Price of Heifers*, 1935–64, offered a new and unexpected problem, encountered also in connection with the Price of Milk. The computations converged rapidly (in 18 iterations) in the frequency domain, but the matrix of second-derivation of the log-likelihood function was not negative definite, as it should have been at a true maximum. On inspection, the trend-cycle component was found to have a non-invertible moving-average part, i.e., a moving-average with roots

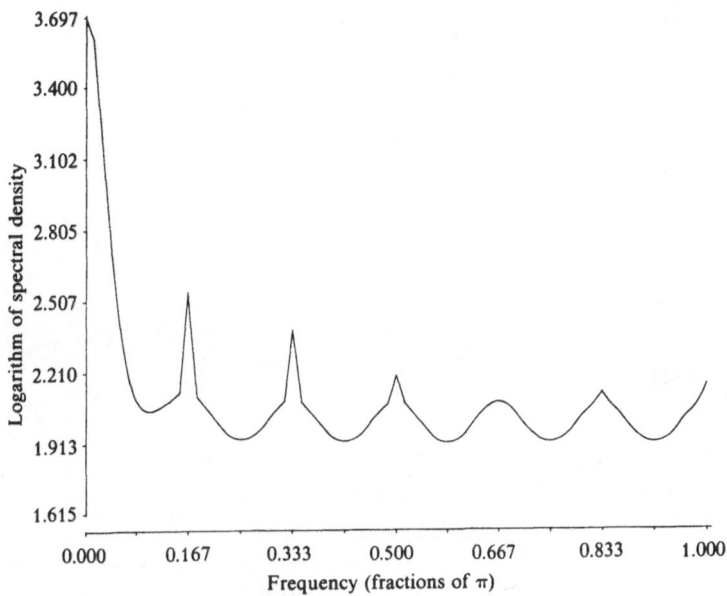

Figure 16: Theoretical spectral density, model for detrended price of cows estimated in the frequency domain, 1925–1964.

inside the unit circle. Since the actual value of the likelihood function is unaffected by whether the estimated roots lie inside or outside the unit circle (a corresponding root always enters the likelihood function symmetrically), the estimation procedure was restarted using as initial values all the coefficients previously obtained except those in the trend-cycle moving-average, which were replaced by the coefficients corresponding to the reciprocal of the root found to be inside the unit circle. Convergence to a new model with a negative-definite matrix of second derivatives of the log-likelihood function, the same value of the function, and all roots well outside the unit circle was obtained in 10 more iterations. Figure 9 shows that the Price of Heifers does not have any marked trends or other important distinguishing characteristics. Figure 17 shows that the theoretical spectral density of the estimated model fits the actual tolerably well: There is a little too much power at the high end of the frequency spectrum and a little too little at the low end.

The *Price of Milk,* 1910–64, presented the same problem as the Price of Heifers: After 21 iterations the procedure converged but with a non-negative matrix of second derivatives of the log-likelihood

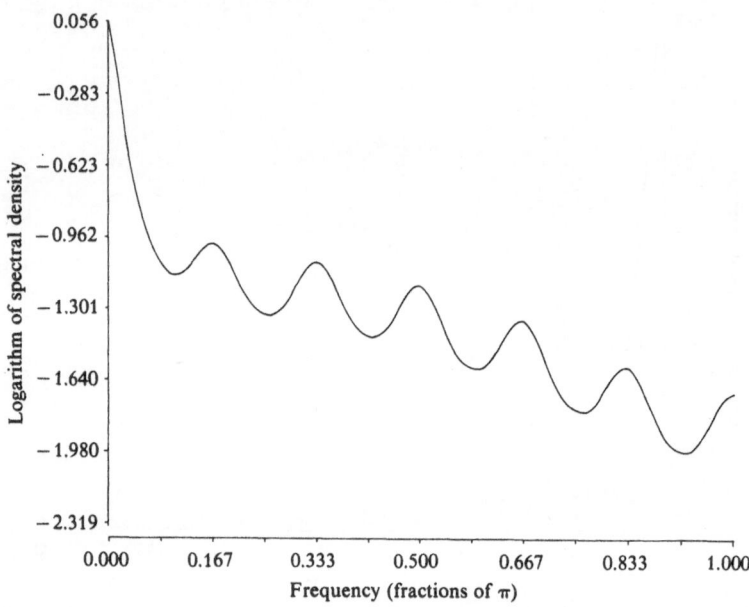

Figure 17: Theoretical spectral density, model for detrended price of heifers estimated in the frequency domain, 1935–1964.

function and a non-invertible moving-average in the trend-cycle component. Restarting with the same modification as outlined above resulted in convergence in 9 more iterations to the estimates presented in the Table. The estimated spectral density for the actual series, shown in Figure 10, suggests a very strong seasonal component at $\pi/6$. This is mirrored in the substantial autoregressive part in the seasonal component of the estimated model with a highly damping moving-average part. This is not sufficient to damp successive spikes at seasonal frequencies to the extent they are damped in the estimated spectral density for the actual series (see Figure 18), but we should remember that the smoothing, i.e., averaging of the estimated values, will have this effect. Again, the problem seems to be too strong an autoregressive part in the trend-cycle component: The theoretical density for the model falls off far too rapidly at very low frequencies.

Cattle Slaughter, 1947–64, represents another type of series in contrast to the four discussed above. Seasonality is very marked and strong in this series as may be seen from the estimated spectral density of the actual series presented in Figure 11. This is revealed by a substantial contribution to the overall variance of the series by the estimated seasonal component, 36.9% for the model estimated in the frequency domain. It was not possible to obtain convergence for a positive σ_3^2. Hence, the irregular component was suppressed. The trend-cycle component has a definite moving-average part although not significantly second-order. The seasonal component is well-defined and damped by a second-order moving-average in its numerator. Comparison of the estimated spectral density of the actual series, Figure 11, with the theoretical spectral density of the fitted model, Figure 19, reveals a moderately good match, albeit one which suggests that the seasonal effect has been dominated by the peaks at high frequencies, $\pi/2$, $2\pi/3$, $5\pi/6$ and π, rather than by the well-defined and marked peak at $\pi/6$.

Hog Slaughter, 1910–64, is similar to Cattle Slaughter in having very well-defined and marked peaks in the spectral density of the actual series at the seasonal frequencies (Figure 12). Although the seasonal component accounted for over 50% of the variance of the observed series, the estimated model fails to capture the marked peaks at $\pi/6$ and $\pi/3$. (See Figure 20.) The trend-cycle component, however, appears to fit quite well.

Industrial Production, 1919–64, shows very marked seasonality all seasonal frequencies except $\pi/6$, where the presence of a seasonal peak may be partly obscured by a strong autoregressive part in the trend-cycle component. (See Figure 13.) The fitted model reflects partly by the importance of the twelfth-order autoregression in the seasonal component but, surprisingly, despite the marked peaks in the theoretical spectral density (see Figure 21), relatively little is contributed by this component to the overall variance in comparison with the model for Hog Slaughter.

A strong second-order autoregression appears in the trend-cycle component and the closeness of one of the roots to the unit circle probably accounts for the high contribution to the overall variance. The spectral density of the fitted model follows the estimated spectral density of the actual series well at frequencies above $\pi/3$, but again drops too quickly at frequencies below this level.

Male Unemployment, 1947–64, shows marked seasonal peaks at $\pi/6$ and $\pi/3$ but little elsewhere. The estimated spectral density for the actual series, as is the case with the other series examined, does not tail off sufficiently rapidly to suggest the absence of a moving-average part in the trend-cycle component. (See Figure 14.) Nonetheless, the fitted model contains a well-defined autoregression in its trend-cycle component but the moving-average is insignificantly different

M. Nerlove

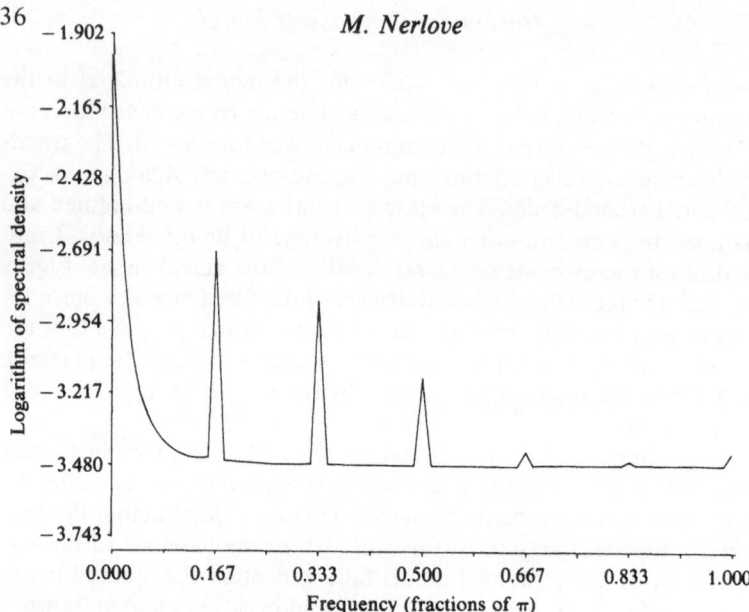

Figure 18: Theoretical spectral density, model for detrended price of milk estimated in the frequency domain, 1910–1964.

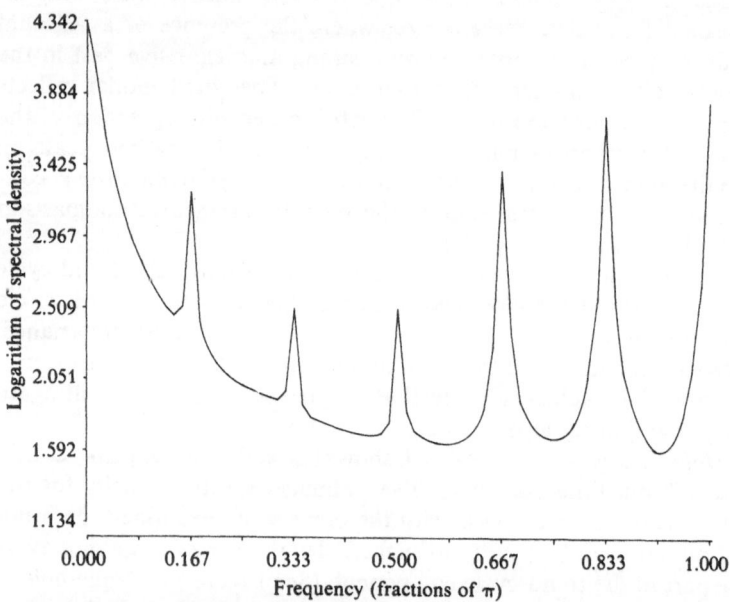

Figure 19: Theoretical spectral density, model for detrended federally inspected cattle slaughter estimated in the frequency domain, 1947–1964.

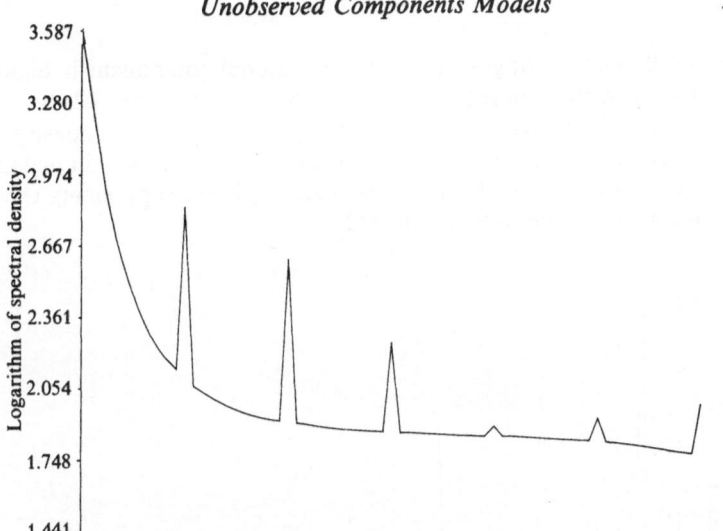

Figure 20: Theoretical spectral density, model for detrended federally inspected hog slaughter estimated in the frequency domain, 1910–1964.

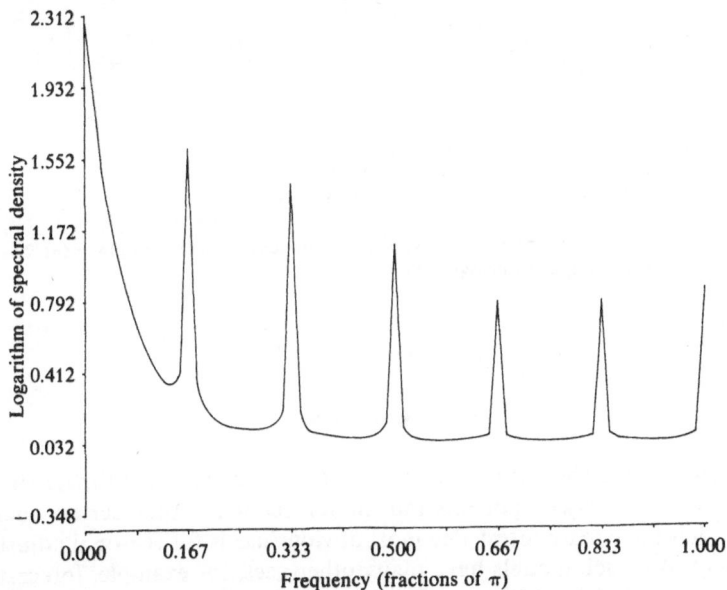

Figure 21: Theoretical spectral density, model for detrended industrial production estimated in the frequency domain, 1919–1964.

38   *M. Nerlove*

from 1.0. The moving-average in the seasonal component is highly significant with complex roots nearly on the unit circle and this accounts for the heavy damping of the theoretical spectral density of the fitted model. (See Figure 22.) The model thus does well in picking up the marked seasonal peak, at $\pi/6$ and $\pi/3$ and suppressing those at all higher frequencies except $\pi/2$.

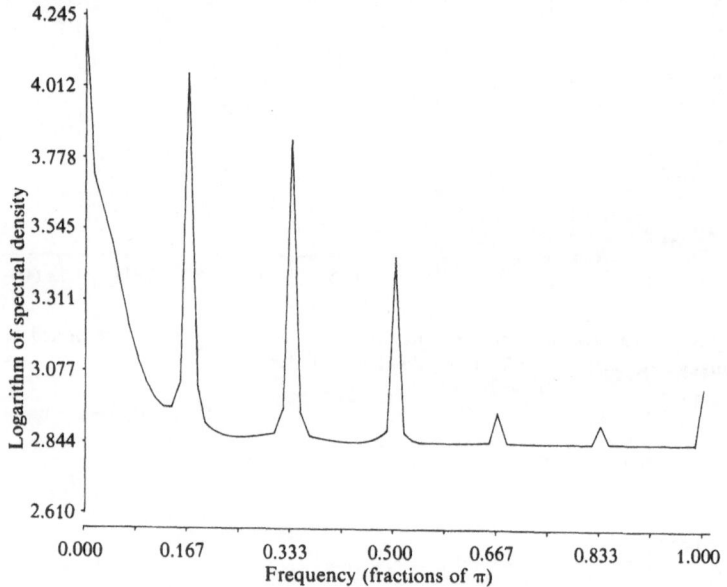

Figure 22: Theoretical spectral density, model for detrended male unemployment, 20+, estimated in the frequency domain, 1947–1964.

5. Conclusion

Our results show that it is possible to formulate and estimate simple unobserved-components models for economic time series. Such models have many uses: The most obvious use is for seasonal adjustment. But such models have many other uses, for example, forecasting, formulation of distributed lag models, and so forth. Many examples are given in Nerlove, *et al.* [9].

Footnotes

[1] ARIMA is used when the series is differenced first before fitting it into the ARMA framework. In general, such differencing is done prior to analysis to render the series stationary. In our work we have more frequently removed a polynomial trend, since differencing is difficult to justify in the context of an unobserved components model. In any case, differencing is equivalent to fitting a polynomial using only the observation at the beginning and end of the series (Durbin [5]). For a discussion of stationarity and other formal properties of time series see, *inter alia*, Nerlove, Grether and Carvalho [9, pp. 22–36]. I do not have time in the course of this lecture to deal adequately with the fundamental concepts of time series analysis such as stationarity and ergodicity.

[2] In general, we also require that the roots of $G(z) = 0$ lie outside the unit circle as well. In this case, the moving-average part of (1.4), $G(U)\varepsilon_t$, is said to be *invertible*, see (Nerlove, *et al.* [9, pp. 40–41]). Invertibility is a desirable property for time series models to have in terms of the uses to which they may be put since invertibility permits us to express all results in terms of the past of the series itself alone.

[3] Under the conditions specified the function $g(z)$ is analytic in an annulus about the unit circle and has a Laurent series expansion there from which the autocovariances of the series $\{x_t\}$ may be recovered as the coefficients of the expansion.

[4] For a discussion of the spectral density function and its estimation see Nerlove, *et al.* [9, pp. 37–68].

[5] In this case $\{u_t\}$ and $\{v_t\}$ are said to be Gaussian.

6. References

[1] Babbage, C. (1856), Analysis of the statistics of the clearing house during the year 1839. *Statistical Journal* XIX.

[2] Brillinger, D. L, (1975), *Time series: data analysis and theory.* New York: Holt, Rinehart and Winston.

[3] Box, G. E. P., and Jenkins, G. M. (1970), *Time series analysis, forecasting and control.* San Francisco: Holden-Day.

[4] Burns, F., and Mitchell, W. C. (1947), *Measuring business cycles.* New York: National Bureau of Economic Research.

[5] Durbin, J. (1961), Trend elimination by moving-average and variate-difference filters. *Bulletin de l'Institut International de Statistique 33rd session,* Paris, 1963.

[6] Jevons, W. (1884), *Investigations in currency and finance.* London: Macmillan.

[7] Koopmans, L. H. (1974), *The spectral analysis of time series.* New York: Academic Press.

[8] Kuznets, S. (1933), *Seasonal variations in industry and trade.* New York: National Bureau of Economic Research.

[9] Nerlove, M., Grether, D. M., and Carvalho, J. L. (1979), *Analysis of economic time series.* New York: Academic Press, Inc.

[10] Persons, W. M. (1919), Indices of general business conditions. *Review of Economic Statistics* I, 5–107.

[11] Pukkila, T. (1977), *Fitting of autoregressive moving average models in the frequency domain.* Report A-6. Department of Mathematical Sciences, University of Tampere. Finnland.

[12] Shiskin, J. (1958), Decomposition of economic time series. *Science* 128, 1539–1546.

[13] Slutzky, E. (1927), The summation of random causes as the source of cyclic processes. *Econometrica* 1937, 5, 105–146. Translated from the earlier paper of the same title in *Problems of Economic Conditions* ed. by the Conjuncture Institute. Moscow.

[14] Yule, G. U. (1921), On the time-correlation problem, with especial reference to the variate-difference correlation method. *Journal of the Royal Statistical Society* 84, 497–526.

[15] Yule, G. U. (1926), Why do we sometimes get nonsense correlations between time series? A study in sampling and the nature of time series. *Journal of the Royal Statistical Society* 89, 1–64.

[16] Yule, G. U. (1927), On a method of investigating periodicities in disturbed series, with special reference to Wolfer's sunspot numbers. *Philosophical Transactions* 226 A, 267–298.

Prediction of Economic Processes with Linear Regression Part

Dortmund

2. The Theoretical Model
3. Model Restrictions for the Residual Process
4. Estimation of the Prediction Function
   4.1 Approximate Maximum Likelihood Estimator
   4.2 Asymptotic Properties of the Estimators
5. An Application
6. References

Die Arbeit behandelt die beste lineare und beste lineare erwar-
tungstreue Vorhersage von Wirtschaftsprozessen mit linearem Regressionsanteil. Dabei muß zum Zweck der Schätzung der Restpro-
zeß geeignet parametrisiert werden. Das Beispiel des autoregressiven
Restes wird dabei ausführlicher behandelt. Bedingungen für Konsistenz und asymptotische Normalverteilung bei der vorgeschlagenen
Prozedur werden angegeben. Das Verfahren wird schließlich ange-
wandt zur nachträglichen Vorhersage der Nettoproduktion und des
Umsatzes im Bekleidungsgewerbe, wobei als Regressoren die Auf-
tragseingänge dienen.

Summary:

For predicting economic processes when additional information by
leading indicators is available, a mixed model with a linear regression
part and a regular residual process is considered. For estimation purposes the residual process has to be parameterized suitably. The case
of an autoregressive residual is investigated in some detail. The sug-
gested procedure is applied to the net production and sales time series
of the German clothing industry, where as a leading indicator the se-

ries of received orders is taken. The performance is evaluated by several ex-post predictions.

1. Introduction

The regression model with lagged exogenous variables or in the multiple case the reduced form of an econometric model is the classical means of prediction in economics. Since methods developed in time series analysis have found application to economic problems, one realises that they produce fairly often better short-term predictions than the classical regression approach. The Wiener filter, the ARMA- or ARIMA-model, the Kalman filter and similar models are indeed flexible instruments, with which one may achieve a very good fit to the individual dynamic structure of many processes. On the other hand regression models often are less flexible.

Nonetheless time series models are considered with scepticism and reservations among many practitioners. One reason for this may be that their application is widely done without using the knowledge founded on economic theory and the specific insight and experience of the practitioner, which otherwise can be built into a regression model. Thus for some scientists time series models are black boxes with the observations being fed in on one side and the predictions are the outcome on the other side. It is true that a certain parameter adjustment ("model identification") is still necessary, but one cannot argue or reason very much about the results. Because of the simpler structure this may be done far better with the results of a regression analysis.

Due to these reasons it seems advantageous to combine the two approaches. Accordingly one may attempt to bring to bear economic theory and practical experience in the model and to equipe it at the same time with the dynamics and flexibility of a time series approach.

There are various ways to achieve this. A special one is presented in the sequel. Regression models generally are suitable to explain the impact of certain influence factors on the process under consideration when the intensity and the direction of the influences are relatively stable in time. Besides these explainable influences there often remain certain dynamical (e.g. short term) influences in the residual process, especially when the observations are made over short intervals. In these case the residuals don't obey the orthogonality assumption of the classical model.

Under not very stringent assumptions some of the desirable properties of the usual regression estimators – such as unbiasedness and consistency (under stronger assumptions also asymptotic normality) – are nonetheless preserved.

But if the residual process still contains movements with some apparent intrinsic regularity of one sort or another, then these suggest themselves for further estimation. The result of this estimation can be utilized for extrapolation purposes.

2. The Theoretical Model

In the sequel we describe a model which gradually will be refined until we obtain for it a linear unbiased predictor in the sense of Goldberger [5]. We will see that we still have to restrict the structure considerably further, if the structural parameters of the considered process (e. g. the autocovariances) are unknown and have to be estimated from the time series. We analyse this in the second section.

We begin with the following two assumptions:

A1: Let x_{ti}, u_t be square integrable random variables on some probability space $(\Omega, \mathfrak{A}, P)$ for $t \in \mathbb{Z}$, $i = 1, \ldots, k$ with $E(u_t) = 0$, $t \in \mathbb{Z}$.

A2: For the observation period $\mathbb{T} := \{1, 2, \ldots, T\}$ and the prediction period $\mathbb{T}^* := \{T+1, \ldots, T+Q\}$, T, $Q \in \mathbb{N}$ the relation

$$(2.1) \qquad y_t = \beta_1 x_{t1} + \beta_2 x_{t2} + \cdots + \beta_k x_{tk} + u_t$$

or

$$(2.2) \qquad y_t = x'(t)\beta + u_t$$

in vector notation holds.

For the observation period we write the model as usual in the form $y = X\beta + u$. $\Sigma := E(uu')$ is the covariance matrix of the residual process in the observation period and for $s \in \mathbb{T}^*$ $\sigma_s := E(u_s u)$ is a T-vector of autocovariances. For simplicity we do not mention explicitly the dimension of vectors and matrices and hope that this will not confuse the reader.

If we denote by y_s^p a predictor for y_s, $s \in \mathbb{T}^*$, then

$$(2.3) \qquad M(y_s^p) := E(y_s^p - y_s)^2$$

is the mean square prediction error. We speak of a *best prediction*, if this mean square error is minimized within the considered class of predictors.

As may be seen immediately from (2.3) by differentiation, the best *linear* prediction $y'c_0 = \sum_{t=1}^{T} c_{0,t} y_t$ for y_s is determined by the vector

(2.4) $c_0 = [Eyy']^+ (Ey_s y)$,

where by A^+ we denote a generalized inverse of A and

(2.5) $E(yy') = E(X\beta\beta' X') + \Sigma + E(X\beta u') + E(u\beta' X')$

(2.6) $E(y_s y) = E(X\beta\beta' x(s)) + \sigma_s + E(u_s X\beta) + E(ux(s)'\beta)$.

If a linear prediction $y'd$ is to be unbiased for y_s, then the side condition $E(y'd) = E(y_s)$ or equivalent $E(d'X\beta) = E(x(s)'\beta)$ has to be observed. Since the latter has to hold for all $\beta \in \mathbb{R}^k$, this is equivalent to the condition

(2.7) $(EX)d' = E(x(s))$.

Taking this condition into account we are led (by using Lagrange multipliers) to the optimal solution

(2.8) $d_0 = (Eyy')^+ (E(y_s y)) + (Eyy')^+ EX$
                      $\cdot [(EX)'(Eyy')^+ EX]^+ (Ex(s)) - (Eyy')^+ EX$
                      $\cdot [(EX)'(Eyy')^+ EX]^+ (EX)'(Eyy')^+ E(y_s y)$.

In the generality of the model maintained so far the two last mixed terms in (2.5) and (2.6) with elements of the kind $E(u_s x(t)'\beta)$, $s, t \in \mathbb{T} \cup \mathbb{T}^*$ don't vanish. Therefore we introduce a further model restriction. Let

$$\mathfrak{A}_x := \sigma(x_{ti}, i = 1, \ldots, k; t = 1, \ldots, T+Q)$$

be the σ-subalgebra of $\mathfrak{A}$ generated by the regressors in $\mathbb{T} \cup \mathbb{T}^*$.

A3: Let $E(u_t \mid \mathfrak{A}_x) = 0$ for all $t \in \mathbb{T} \cup \mathbb{T}^*$.

Under A1 to A3 the expressions Eyy' and $Ey_s y$ for the best linear resp. the best linear unbiased predictions in (2.4) and (2.8) have the form

(2.9) $Eyy' = E(X\beta\beta' X') + \Sigma$,

(2.10) $Ey_s y = E(X\beta\beta' x(s)) + \sigma_s$.

With the restriction A3 the model is still unsatisfactory for practical purposes. In the models which may be found in the literature in this context therefore the regressors are looked at as exogenous *nonstochastic* variables. Under this assumption and for Σ regular Goldberger derived in [5] the best linear unbiased predictor. In a very general framework, but also with nonstochastic regressors the best linear and the best linear unbiased prediction for processes with a linear regression part is treated by Läuter in [14], [15] and [16]. Läuter considers the regular and the singular case. He also takes into account the exploitation of additional information.

We do not assume in the sequel that additional information is available and stipulate now within our model the assumption

A4: The regressors x_{ti}, $t = 1, \ldots, T + Q$; $i = 1, \ldots, k$ are known nonstochastic quantities.

Under this setup regressors may be included repeatedly in lagged form. Together with economic extrapolation problems regressors often will be leading indicators.

We remark that instead of A4 we could get along with the slightly weaker assumption to restrict our model to the probability space $(\Omega, \mathfrak{A}_x, P(\cdot \,|\, \mathfrak{A}_x))$, where $P(\cdot \,|\, \mathfrak{A}_x)$ denotes the conditional probability subject to $\mathfrak{A}_x$. For further discussion we drop this point of view, since it doesn't lead to any real insights for practical purposes.

With the preceding considerations we arrive at the following result:

Lemma 1: *Under the assumptions* A1, A2 *and* A4 *there holds:*
I. There always exists a best linear predictor $y' c_0$ *for* y_s, $s \in \mathbb{T}^*$, *and* c_0 *is given by*

$$(2.11) \qquad c_0 = [X\beta\beta' X' + \Sigma]^+ (X\beta\beta' x(s) + \sigma_s) \,.$$

II. Best linear unbiased predictors $y' d_0$ *for* y_s, $s \in \mathbb{T}^*$ *exist, if and only if there are vectors* $d \in \mathbb{R}^T$ *with* $X' d = x(s)$. *In this case* d_0 *is given by*

$$(2.12) \qquad \begin{aligned} d_0 &= \Sigma^+ X(X' \Sigma^+ X)^+ x(s) \\ &\quad + \Sigma^+ \sigma_s - \Sigma^+ X(X' \Sigma^+ X)^+ X' \Sigma^+ \sigma_s \,. \end{aligned}$$

The result (2.12) may be derived simply by inserting $x(s)$ for $X' d$ in (2.3) according to the side condition and then applying the Lagrange approach.

It should be noted in the above Lemma that the best linear prediction depends on the parameter vector β, whereas the best linear unbiased prediction does not. Besides this the latter result corresponds to a quite intuitive procedure. Let us assume the process (u_t) were observable. Then the best linear prediction for u_s, $s \in \mathbb{T}^*$, given u would be

(2.13) $u_s^p = \sigma_s \Sigma^+ u$.

If the covariances are known, then under A1, A2 and A4 it suggest, almost itself to first estimate β by

(2.14) $\tilde{\beta} = [X' \Sigma^+ X]^+ X' \Sigma^+ y$,

to evaluate then the regression residuals

$$\tilde{u} = y - X \tilde{\beta}$$

and to apply the extrapolation formula (2.13) to these residuals,

(2.15) $\tilde{u}_s^p = \sigma_s \Sigma^+ \tilde{u}$.

This leads to the composed prediction

$$y_s^p = x'(s) \tilde{\beta} + \tilde{u}_s^p .$$

Inserting everything we arrive at $y_s^p = y' d_0$ with d_0 given by (2.12).

3. Model Restrictions for the Residual Process

To get extrapolation formulae which do work in practical cases we introduced in the second section model restrictions concerning the regression part. The structure of the residual process was allowed to be fairly general.

The application of the results of Lemma 1 requires for the best linear prediction the knowledge of the parameter vector β and of the covariances of the residual process. For the best linear unbiased prediction the covariances have to be known. In practical applications both β and the covariances are usually unknown.

To be able to derive reasonable estimators of the covariances from the observations we now have to restrict suitably the structure of the residual process. For asymptotic considerations (consistency, asymptotic normality) we finally have to introduce some assumptions concerning the asymptotic behaviour of the regressors.

Estimators for the covariances that are mean square consistent can be derived, if the residual process is stationary up to order four and covariance ergodic. If in this case we introduce the notation

$$\sigma(t - s) := \sigma_{ts}$$

for the covariances of the residuals, then

$$\hat{\sigma}(h) := \frac{1}{T} \sum_{t=h+1}^{T} u_t u_{t-h} \quad (h \geq 0)$$

is a nonnegative definite function converging to $\sigma(h)$ in the mean.

The above assumption is still too general, since for estimating the $T \times T$-matrix Σ the elements $\sigma(0)$, $\sigma(1)$, ..., $\sigma(T-1)$ would have to be estimated from T observations. A practical rule of thumb says that one should not go beyond a maximal lag of about $T/3$ by estimating autocovariances from T observations.

Thus the residual process has to be parameterized suitably. A fairly general formulation of a parameterized mixed linear model is the ARMA scheme

$$y_t = \alpha_1 y_{t-1} + \cdots + \alpha_r y_{t-r} + \beta_1 x_{t1} + \cdots + \beta_k x_{tk} + u_t$$

$$u_t = \gamma_1 u_{t-1} + \cdots + \gamma_p u_{t-p} + \varepsilon_t + \delta_1 \varepsilon_{t-1} + \cdots + \delta_q \varepsilon_{t-q}$$

and

$$\varepsilon \in L_2^{\mathbb{Z}}(\Omega, \mathfrak{A}, P) \text{ is white noise.}$$

If we denote by $\mathcal{L}$ the lag operator and introduce the operator polynomials

$$A(\mathcal{L}) = I - \alpha_1 \mathcal{L} - \cdots - \alpha_r \mathcal{L}^r$$

$$C(\mathcal{L}) = I - \gamma_1 \mathcal{L} - \cdots - \gamma_p \mathcal{L}^p$$

$$D(\mathcal{L}) = I + \delta_1 \mathcal{L} + \cdots + \delta_q \mathcal{L}^q$$

(I is the identity operator), then the above model may be written in the form

$$(3.1) \qquad A(\mathcal{L}) y_t = \sum_{i=1}^{k} \beta_i x_{ti} + u_t$$

$$(3.2) \qquad C(\mathcal{L}) u_t = D(\mathcal{L}) \varepsilon_t$$

or

$$(3.3) \qquad C(\mathcal{L}) A(\mathcal{L}) y_t = \sum_{i=1}^{k} \beta_i C(\mathcal{L}) x_{ti} + D(\mathcal{L}) \varepsilon_t.$$

If the inverse operator of $C(\mathcal{L})$ exists, then we finally get

$$A(\mathcal{L}) y_t = \sum_{i=1}^{k} \beta_i x_{ti} + C(\mathcal{L})^{-1} D(\mathcal{L}) \varepsilon_t.$$

Several recent contributions treat the parameter estimation within such a model or (mostly) for the one or the other special case (e.g. $A(\mathscr{L}) = I$, $C(\mathscr{L}) = I$ or $D(\mathscr{L}) = I$). If a lag relation in the regression part is stated explicitly, for example in the form

$$\sum_{l=0}^{n} \sum_{i=1}^{k} \beta_i x_{t-l,i},$$

then also the transfer function models ([3], chap. 10) belong to this class of models. A certain review may be found e.g. in the recent paper of Reinsel [19]. He derives maximum likelihood estimators for the model (3.3) for the case of normally distributed ε_t.

With respect to our aim and to the applications in the last section we also will confine us here to certain special cases of the general model.

Firstly we stipulate $A(\mathscr{L}) = I$. That means we don't consider an autoregressive component. This corresponds to the introductory remarks, where we mentioned that we want to include the fluctuations of the process after subtraction of the regression part into the residual part. Thus cases in which pertinent reasons imply an autoregressive component are not covered by our considerations. Another reason for introducing an autoregressive component can be the modelling of seasonal fluctuations. In this case $A(\mathscr{L})$ has the form $A(\mathscr{L}) = A_1(\mathscr{L}) \cdot A_2(\mathscr{L}^s)$, where s is the period of the season (see [3], chap. 9). If seasonal fluctuations can be explained – at least partly – by regression (e.g. by seasonal dummies), then there does not seem to be an urgent need for an AR-part in the model.

With $A(\mathscr{L}) = I$ we have a process with a linear regression part and an ARMA-residual. (It is thus neither an AREX- nor an ARMAX-type model.)

In view of the applications in the fifth section the model has further been simplified by putting $D(\mathscr{L}) = I$. The advantage of this simplification is that numerical estimation algorithms are simplified and that they correspond to the intuitive procedure mentioned in the second section: β is estimated according to (2.14) by a two-stage quasi-Aitken estimation and an extrapolation formula of the type (2.15) is applied to the regression residuals. Of course this simplification also has certain disadvantages. Modelling the residual process only by an autoregressive scheme requires the estimation of a large number of parameters. It can't be excluded therefore that by adding an MA-part and reducing the number of parameters better predictions (in the mean square sense) may be attained.

We specify our preceding considerations in the following assumption

A5:

$$(3.4) \qquad u_t = \sum_{j=1}^{p} \gamma_j u_{t-j} + \varepsilon_t \,,$$

(ε_t), $t \in \mathbb{Z}$ is a sequence of independent random variables on $(\Omega, \mathfrak{A}, P)$ with $E(\varepsilon_t) = 0$, $E(\varepsilon_t^2) = \sigma^2 > 0$ and the ε_t are either identically distributed or $E|\varepsilon_t|^{2+\varepsilon} < K$ holds for certain constants $\varepsilon > 0$ and $K > 0$ and all $t \in \mathbb{Z}$. Moreover all zeroes of the characteristic polynomial

$$(3.5) \qquad z^p - \gamma_1 z^{p-1} - \cdots - \gamma_p \,, \qquad z \in \mathbb{C}$$

are inside the unit disk.

4. Estimation of the Prediction Function

4.1 Approximate Maximum Likelihood Estimator

With the parameterization of the residual process of the last section our model is now described by the assumptions A2, A4 and A5. In order to derive approximate maximum likelihood estimators for the parameters of the best linear unbiased prediction we assume for simplicity that the process is also observable at the instants $-p+1$, $-p+2, \ldots, 0$. Besides this we slightly sharpen the above model assumptions by stipulating that the residual process equals zero at these instants (model with fixed initial conditions). The effect of this restriction is asymptotically negligible and we shall drop it for asymptotic considerations. If however T is not very large and if the zeroes of the polynomial (3.5) are near the unit circle – the damping thus going slowly – this restriction may play a non-negligible part.

For the derivation of a ML-estimator we further stipulate

A6: ε_t is normally distributed for each $t \in \mathbb{T} \cup \mathbb{T}^*$.

For

$$z_* := (z_{1-p}, \ldots, z_0, z_1, \ldots, z_T)' \in \mathbb{R}_{T+p} \,,$$

$$z := (z_1, \ldots, z_T)' \in \mathbb{R}_T \,.$$

Let L^j, $j = 0, 1, \ldots, p$ be the shift

$$L^j z = (z_{1-j}, z_{2-j}, \ldots, z_{T-j})' \,.$$

If we introduce the polynomial $C(L)$ corresponding to $C(\mathcal{L})$ in section three, then we have $C(L)u = \varepsilon$ with $u = y - X\beta$. It thus follows from the above assumptions that the joint density of y equals

$$(2\pi)^{-T/2}(\sigma^2)^{-T/2}\exp\cdot\left\{-\frac{1}{2\sigma^2}[C(L)(y-X\beta)]'[C(L)(y-X\beta)]\right\}.$$

From this we get the log-likelihood function

$$\log\phi(\beta,\gamma,\sigma^2) = -\frac{T}{2}\log 2\pi - \frac{T}{2}\log\sigma^2$$

$$-\frac{1}{2\sigma^2}[C(L)(y-X\beta)]'\cdot[C(L)(y-X\beta)]$$

with the partial derivatives

(4.1)
$$\frac{\partial\log\phi}{\partial\beta'} = -\frac{1}{\sigma^2}[C(L)X]'[C(L)X]\beta$$

$$+\frac{1}{\sigma^2}[C(L)X]'[C(L)y],$$

(4.2)
$$\frac{\partial\log\phi}{\partial\gamma'} = -\frac{1}{\sigma^2}\frac{\partial C(L)(y-X\beta)}{\partial\gamma'}C(L)(y-X\beta)$$

$$=\frac{1}{\sigma^2}[L(y-X\beta),\ldots,L^p(y-X\beta)]'$$

$$\cdot[C(L)(y-X\beta)],\qquad\gamma:=(\gamma_1,\ldots,\gamma_p)',$$

(4.3)
$$\frac{\partial\log\phi}{\partial\sigma^2} = -\frac{T}{2\sigma^2} + \frac{1}{2\sigma^4}[C(L)(y-X\beta)]'$$

$$\cdot[C(L)(y-X\beta)].$$

This leads us to the ML-equations

(4.4)
$$[C(L)X]'[C(L)X]\beta = [C(L)X]'[C(L)y]$$

(4.5)
$$([L^i(y-X\beta)]'[L^j(y-X\beta)])_{i,j=1,\ldots,p}\cdot\gamma$$

$$=([L^i(y-X\beta)]'(y-X\beta))_{i=1,\ldots,p}$$

(4.6)
$$\sigma^2 = \frac{1}{T}[C(L)(y-X\beta)]'[C(L)(y-X\beta)].$$

In (4.4)–(4.6) β, γ, σ^2 are variables. Elsewhere they are the fixed, but unknown parameters of the model. (We hope that this will not confuse the reader.) The solutions of (4.4)–(4.6) will be denoted by β^*, γ^* and σ^{*2}, respectively.

Equation (4.4) corresponds to an Aitken estimation for β, since $C(L)y = C(L)X\beta + C(L)u = C(L)X\beta + \varepsilon$ with $E(\varepsilon\varepsilon') = \sigma^2 I$. The equations (4.5) are the normal equations for the autoregressive parameters of the residual process. They correspond to the estimation form of the so-called Yule-Walker-equations.

The above results suggest the following simple practical procedure: To begin with a first estimation of the vector β (e. g. by ordinary least squares) is effectuated. Then the regression residuals of this first estimation are used in (4.5) to get a first estimation of γ. This latter estimation is used in the second stage in (4.4) for a quasi-Aitken estimator of β. The procedure can be continued iteratively. (In all applications up to now the procedure always converged, and already after a few steps three or four digits of the results coincided.)

4.2 Asymptotic Properties of the Estimators

One can't say much about the finite properties of the above procedure. Of course the resulting predictions are not best linear unbiased predictions in the sense of section two. Therefore we will briefly discuss the asymptotic properties. For this we drop the assumptions stipulated at the beginning of this section (fixed initial conditions and normality of the ε_t). The asymptotic properties depend on the consistency of the first estimators for β and γ.

Lemma 2: *A sufficient condition for the mean square consistency of the ordinary least squares estimator $\hat{\beta}$ for β in the first stage under the assumptions A2, A4 and A5 is, that*

$$(4.7) \qquad \lambda_{\min}(X'X) \to \infty \quad for \quad T \to \infty .$$

In (4.7) $\lambda_{\min}$ is the smallest eigenvalue of the matrix $X'X$, which of course depends on T (we don't express this explicitly and hope that this simplification will not confuse).

Proof: Firstly it follows from (4.7) that $X'X$ is invertible for almost all T. Thus $\hat{\beta} = (X'X)^{-1}X'y$ and $\hat{\beta} - \beta = (X'X)^{-1}X'u$ are well defined for almost all T. $\hat{\beta} - \beta$ has expectation $\mathbf{0}$ and covariance matrix

$$V(\hat{\beta} - \beta) = (X'X)^{-1}X'\Sigma X(X'X)^{-1} .$$

To obtain an upper bound for V we use a technique which was very successfully introduced by Willers in [22]. For the maximal eigenvalue of V we have

$$\lambda_{\max} V(\hat{\beta} - \beta)$$
$$\leq \lambda_{\max}^2 [(X'X)^{-1/2}] \lambda_{\max} [(X'X)^{-1/2} X'] \lambda_{\max} [\Sigma] .$$

With

$$\lambda_{\max}^2 [(X'X)^{-1/2} X']$$
$$:= \lambda_{\max} [(X'X)^{-1/2} X' X(X'X)^{-1/2}] = 1$$

and

$$\lambda_{\max}^2 [(X'X)^{-1/2}] = \lambda_{\min}^{-1} (X'X)$$

we get

$$\lambda_{\max} V(\hat{\beta} - \beta) \leq \frac{\lambda_{\max} \Sigma}{\lambda_{\min} X'X} .$$

From assumption A5 it follows that the maximal eigenvalue of the $(T \times T)$-matrix Σ is bounded for all T. Thus (4.7) implies $V(\hat{\beta} - \beta) \to 0$ for $T \to \infty$.

Lemma 3: *If in the second step the OLS-residuals $\hat{u} = y - X\hat{\beta}$ are used in the* Yule-Walker *equations to estimate $\sigma(h)$ by*

$$\hat{\sigma}(h) = \frac{1}{T} \sum_{t=1}^{T} \hat{u}_t \hat{u}_{t-h}, \qquad h = 0, 1, \ldots, p ,$$

then A5 *and*

(4.8) $\lambda_{\min} X'X > 0$ *for one $T < \infty$*

are sufficient for the weak consistency of $\hat{\sigma}(h)$ and of $\hat{\gamma}$.

Proof: If $\lambda_{\min} X'X > 0$ for T^*, then $X'X$ is invertible for all $T \geq T^*$. Thus $\hat{u} = u - X(X'X)^{-1} X'u$ is well defined for almost all T. We take the estimator $\hat{\sigma}(0) = \frac{1}{T} \hat{u}' \hat{u}$. (For the autocovariances one can reason correspondingly by taking account of the shifts.) We have

$$\frac{1}{T} \hat{u}' \hat{u} = \frac{1}{T} u' u - \frac{1}{T} u' X(X'X)^{-1} X'u .$$

The first term on the right hand side converges under A5 stochastically to $\sigma(0)$ (see e.g. [1], p. 195). With

$$z := \frac{1}{\sqrt{T}} (X'X)^{-1/2} X'u$$

the second term is given by $z'z$. We obviously have $E(z) = 0$ and

$$V(z) = \frac{1}{T} (X'X)^{-1/2} X' \Sigma X (X'X)^{-1/2}.$$

Corresponding to the proof of Lemma 2 we get an upper bound for $V(z)$ by

$$\lambda_{max} V(z) \le \frac{1}{T} \lambda_{max}^2 [(X'X)^{-1/2} X'] \lambda_{max} \Sigma \le \frac{1}{T} \lambda_{max} \Sigma.$$

Thus $V(z) \to 0$ for $T \to \infty$.

The consistency of the estimator $\hat{\gamma}$ follows from the consistency of the $\hat{\sigma}(h)$ by Slutsky's theorem.

We now give sufficient conditions for the asymptotic normality of the solution of (4.4) and (4.5). For this we sharpen assumption A5 and condition (4.7):

A7: The random variables ε_t, $t \in \mathbb{Z}$ are identically distributed.

A8: The limits $M_{ij} := \lim_{T \to \infty} \frac{1}{T} (L^i X)'(L^j X)$, $i, j = 0, \ldots, p$ exist and M_{00} is regular.

Remark: Assumption A7 can presumably be replaced by a suitable condition of uniform integrability (see Eicker [4]) and assumption A8 can presumably be replaced by a weaker condition of the kind $\lim_{T \to \infty} \max_{1 \le i \le k} x(i)'(X'X)^{-1} x(i) = 0$. But we don't go into a further discussion of these points here.

Let β^* be a generalized least squares estimator of β, i.e. a solution of equation (4.4), which we write in the form

$$M_T \beta_T^* = \frac{1}{T} [C(L)X]' C(L)y,$$

where

$$M_T := \frac{1}{T} [C(L)X]' [C(L)X].$$

Because of $C(L)u = \varepsilon$ we have

$$M_T \sqrt{T}(\beta^* - \beta) = \frac{1}{\sqrt{T}} [C(L)X]' \varepsilon =: m_T.$$

It follows from A8 that M_T converges for $T \to \infty$ to a regular limit, which we denote by M. It is well known that under A7 the right hand side is asymptotically normally distributed with mean vector zero and covariance matrix $\sigma^2 M$ (see e.g. [20], p. 183/184). If in the second (and in further) stages of the estimation procedure in $C(L)$ in equation (4.4) consistent estimators for γ are inserted then it follows for the resulting solution $\tilde{\beta}_T$, that

$$\plim_{T \to \infty} \sqrt{T}(\tilde{\beta}_T - \beta_T^*) = 0.$$

This may be seen e.g. by transfering the proof given in [20] (p. 211–213) to the above case, which may be done very easily. It thus follows that $\tilde{\beta}_T$ has the same asymptotic distribution as β_T^*. We stated already that subject to A5

$$\sigma^*(i-j) = \frac{1}{T}(L^i u)'(L^j u)$$

or the estimator

$$\frac{1}{T} \sum_{t=1}^{T} u_t u_{t-|i-j|},$$

which is asymptotically equivalent to $\sigma^*(i-j)$, is a consistent estimator for $\sigma(h)$. Also, the estimator of the variance of the innovation process, derived from (4.6),

$$\sigma_T^{*2} = \frac{1}{T}[C(L)u]'[C(L)u] = \frac{1}{T}\varepsilon'\varepsilon$$

is consistent for σ^2.

Let

$$\sigma := (\sigma(1), \ldots, \sigma(p))', \qquad \sigma_T^* := (\sigma^*(1), \ldots, \sigma^*(p))'$$

and

$$\Sigma(p) := (\sigma(i-j))_{i,j=1,\ldots,p},$$
$$\Sigma_T^*(p) := (\sigma_T^*(i-j))_{i,j=1,\ldots,p}.$$

We consider the estimator γ_T^*, derived from (4.5);

$$\Sigma_T^*(p)\gamma_T^* = \sigma_T^*.$$

From the model relation $C(L)u = \varepsilon$ we obtain

$$\Sigma_T^*(p)(\gamma - \gamma_T^*) = \frac{1}{\sqrt{T}} s_T$$

where

$$s_T = \left(\frac{1}{\sqrt{T}} \varepsilon' L^i u\right)_{i=1,\ldots,p}.$$

From A5 it follows that $E(s_T) = 0$ and $V(s_T) = \sigma^2 \Sigma(p)$. Since $\Sigma_T^*(p) \to \Sigma(p)$ and $\sigma_T^* \to \sigma$ stochastically with $T \to \infty$, γ_T^* is consistent for γ. The addition of A7 gives a sufficient condition for s_T being asymptotically normally distributed with mean **0** and covariance matrix $\sigma^2 \Sigma(p)$.

A proof for this statement may be found e.g. in [1] (p. 200 ff. and p. 427 ff.). In the proof given there it is shown that any linear combination of the elements of s_T,

$$b' s_T = \frac{1}{\sqrt{T}} \sum_{j=1}^{p} b_j(\varepsilon' L^j u), \qquad b \in \mathbb{R}^p \text{ arbitrary}$$

is asymptotically normally distributed. Let $a \in \mathbb{R}^k$ be arbitrarily chosen. Then

(4.9) $$a' m_T + b' s_T = \frac{1}{\sqrt{T}} \varepsilon' \left(c + \sum_{j=1}^{p} b_j L^j u\right)$$

with

$$c := C(L) X a$$

is an arbitrary linear combination of m_T and s_T. By transfering the cited argumentation, given by T. W. Anderson, line by line to (4.9) we get the result that (4.9) also is asymptotically normally distributed with mean 0 and variance $\sigma^2 a' M a + \sigma^2 b' \Sigma(p) b$. Thus m_T and s_T are asymptotically jointly normally distributed and independent. It follows that

$$\sqrt{T} \begin{pmatrix} \beta_T^* - \beta \\ \gamma_T^* - \gamma \end{pmatrix} \text{ has the asymptotic distribution}$$

$$N\left(\mathbf{0}, \sigma^2 \begin{pmatrix} M^{-1} & 0 \\ 0 & \Sigma^{-1}(p) \end{pmatrix}\right).$$

It remains to show that subject to A4, A5, A7 and A8 the same asymptotic results hold, if by estimating $\sigma(h)$ the unobservable process (u_t) is replaced by the regression residuals, where β is consistently estimated by $\tilde{\beta}_T$.

Let $\tilde{\sigma}_T(i-j)$, $\tilde{\sigma}_T$, $\tilde{\Sigma}_T(p)$ be the corresponding estimators, where u is replaced by $\tilde{u} = y - X\tilde{\beta}_T = u - X(\tilde{\beta}_T - \beta)$.

With this we have

$$\tilde{\sigma}_T(i-j) = \frac{1}{T}(L^i\tilde{u})'(L^j\tilde{u})$$

$$= \frac{1}{T}[L^i(u-X(\tilde{\beta}_T-\beta))]'[L^j(u-X(\tilde{\beta}_T-\beta))]$$

$$= \sigma_T^*(i-j) + (\tilde{\beta}_T-\beta)'\frac{1}{T}(L^iX)'(L^jX)(\tilde{\beta}_T-\beta)$$

$$- \frac{1}{T}(L^iu)'[L^jX(\tilde{\beta}_T-\beta)] - \frac{1}{T}[L^i(X(\tilde{\beta}_T-\beta))]'L^ju$$

or

$$\sqrt{T}(\tilde{\sigma}_T(i-j) - \sigma_T^*(i-j)) = \sqrt{T}(\tilde{\beta}_T-\beta)\frac{1}{T}(L^iX)'(L^jX)(\tilde{\beta}_T-\beta)$$

$$- \frac{1}{\sqrt{T}}(L^iu)'(L^jX)(\tilde{\beta}_T-\beta)$$

$$- \frac{1}{\sqrt{T}}(L^ju)'(L^iX)(\tilde{\beta}_T-\beta).$$

In the last term (correspondingly in the second) on the right hand side the vector $\frac{1}{\sqrt{T}}(L^ju)'(L^iX)$ has expectation zero and covariance matrix $\frac{1}{T}(L^iX)'\Sigma(L^jX)$. It follows from A5 and A8 that this matrix is bounded for $T\to\infty$. Since further $\frac{1}{T}(L^iX)'(L^jX)$ converges to M_{ij} (assumption A8), since $\sqrt{T}(\tilde{\beta}_T-\beta)$ converges in distribution and since $\tilde{\beta}_T-\beta$ converges to zero stochastically, it follows that

$$(4.10) \qquad \plim_{T\to\infty} \sqrt{T}[\tilde{\sigma}_T(i-j) - \sigma_T^*(i-j)] = 0.$$

This implies $\plim_{T\to\infty} \sqrt{T}(\tilde{\gamma}_T - \gamma_T^*) = 0$, because from

$$\tilde{\Sigma}_T(p)\tilde{\gamma}_T = \tilde{\sigma}_T \quad \text{and} \quad \Sigma_T^*(p)\gamma_T^* = \sigma_T^*$$

we have

$$\hat{\Sigma}_T(p) \sqrt{T}(\tilde{\gamma}_T - \gamma_T^*) + \sqrt{T}[\hat{\Sigma}_T(p) - \Sigma_T^*(p)]\,\gamma_T$$
$$= \sqrt{T}(\tilde{\sigma}_T - \sigma_T^*)\,.$$

From (4.10) we conclude that the right hand side and $\sqrt{T}[\hat{\Sigma}_T(p) - \Sigma_T(p)]$ converge stochastically to a null vector resp. to a null matrix. $\plim_{T \to \infty} \tilde{\gamma}_T = \gamma$ and $\plim_{T \to \infty} \hat{\Sigma}_T(p) = \Sigma(p)$ thus imply

$$\plim \sqrt{T}(\tilde{\gamma}_T - \gamma_T^*) = 0\,.$$

Therefore $\tilde{\gamma}_T$ has the same asymptotic distribution as γ_T^*. We summarize the above results in the following

Theorem: *Under the assumptions* A2, A4, A5, A7 *and* A8 $\sqrt{T}\begin{pmatrix} \tilde{\beta}_T - \beta \\ \tilde{\gamma}_T - \gamma \end{pmatrix}$

is asymptotically normally distributed with mean vector **0** *and covariance matrix* $\sigma^2 \begin{pmatrix} M^{-1} & 0 \\ 0 & \Sigma^{-1}(p) \end{pmatrix}.$

The asymptotic covariance matrix may be estimated consistently by inserting the corresponding estimators $\tilde{\sigma}^2$, $\tilde{M} = \dfrac{1}{T}[\check{C}(L)X]'[\check{C}(L)X]$ and $\hat{\Sigma}_T(p)$.

5. An Application

The procedure described in section 4.1 is worked out in FOR-TRAN IV. The program contains various options, e. g. seasonal dummies, arbitrary lag structure etc. As already mentioned, it converged in all applications hitherto after a few steps. In the sequel for demonstration purposes we select an example of short-term business forecasting. As a special branch the clothing industry is considered. In this branch fairs take place in Düsseldorf, Munich and Berlin in the spring and in the autumn. The bulk of orders is placed by the trading companies during these fairs, in the spring for the autumn and winter collection, in the autumn for the spring and summer collection. Orders placed apart from these fairs are only of minor importance. The monthly series of the index of received orders thus shows extremely strong seasonal variations.

The production starts only after the orders have arrived. The delivery of the production is made about half a year after the orders have come in. Therefore the series of incoming orders has a lead of five to seven months compared to the net production series. Net production and sales are spread over several periods. Therefore these two time series have less extreme seasonal peaks than the incoming orders.

A comparison of the two series indicates that the index of sales shows seasonal variations which are more intensive and increasing in time. The latter fact would recommend a suitable transformation (e.g. a Box-Cox transformation) of this series. (We don't carry out such a transformation in the following application of the procedure.) The seasonal variation could be associated to the residual process.

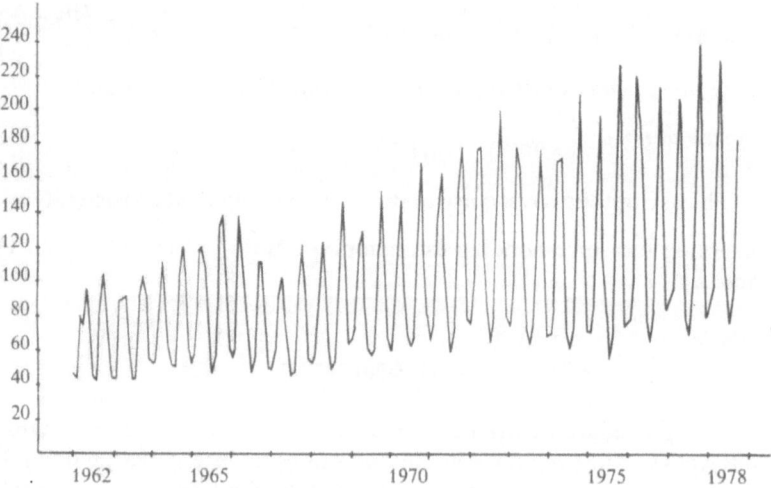

Figure 1: Monthly index of received orders in clothing industry.

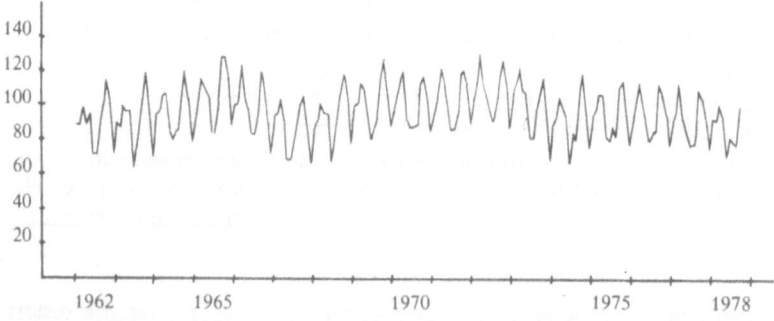

Figure 2: Monthly index of net production in clothing industry.

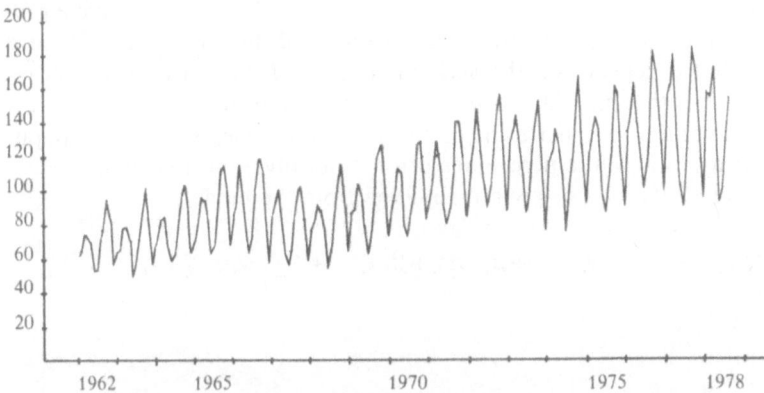

Figure 3: Monthly index of sales in clothing industry.

But in the majority of the applications it turned out to be better to estimate a mean seasonal pattern by dummy variables and to associate only the deviations of this mean pattern to the residual process. Such a procedure implies of course that the seasonal pattern is fairly stable in the course of time. Some of the deviations between predicted and observed seasonal peaks indeed may be caused by shifts of dates of fairs into neighbouring months. (By a refined model such deviations could be avoided – at least partly.)

Because of the seasonal variations the order of the autoregressive scheme in the residual process should not be smaller than 12. An increase of the order beyond this level very often led to only small improvements of the predictions (the orders 18 and 24 were attempted). In several cases the results even deteriorated.

To compare several parameter constellations for a sequence of arbitrary "endpoints of the observation period" T, ex-post predictions were made for the period $\{T+1, \ldots, T+Q\}$. Of course the prediction period was excluded from the parameter estimation. Since also only regressors known up to time T were to be taken into account, the length Q of the prediction period could not exceed the minimal time lag in the model (which is six in all examples presented here).

As goodness criterion for the predictions Theil's inequality coefficient

(5.1)
$$U = \left(\frac{\sum_{q=1}^{Q} (y_{T+q} - y_{T+q}^p)^2}{\sum_{q=1}^{Q} (y_{T+q} - y_T)^2} \right)^{1/2}$$

was used. It compares the actual prediction y^p_{T+q} with the naive prediction $y^p_{T+q} := y_T$. A certain disadvantage of this coefficient of course is that it depends on the accidental value of the last observation y_T. On the other hand it allows a very intuitive interpretation. Secondly it is invariant to linear transformations in the scale, whereas the popular mean absolute percentage error is not invariant to shifts.

The examples below are all taken from the model

$$(5.2) \qquad y_t = \beta_0 + \beta_1 x_{t-6} + \beta_2 x_{t-7} + \sum_{h=1}^{12} \gamma_h u_{t-h} + \varepsilon_t .$$

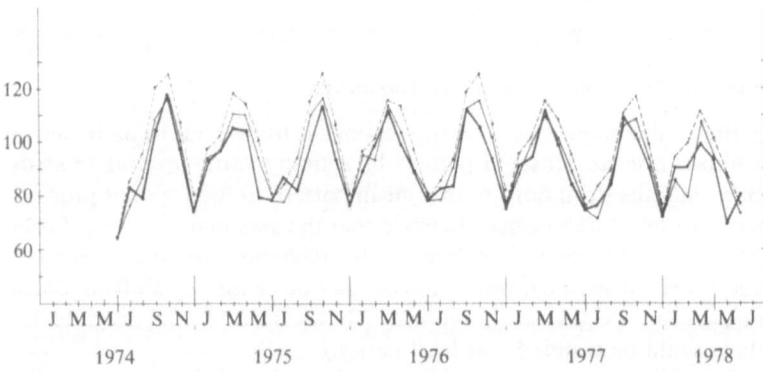

Figure 4: Monthly index of net production, clothing industry; ———, Index of net production; - - - -, Prediction by ordinary least squares; ———, Composed prediction by model (5.2).

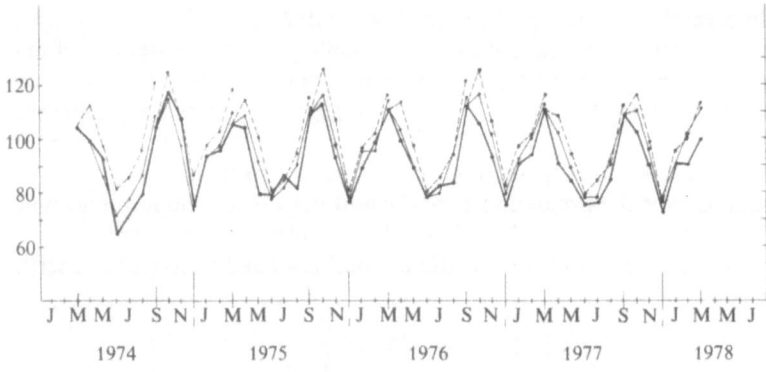

Figure 5: Monthly index of net production, clothing industry; ———, Index of net production; - - - -, Prediction by ordinary least squares; ———, Composed prediction by model (5.2).

Here (x_t) is the monthly series of the index of incoming orders and for (y_t) the series of the index of net production and of sales, respectively, are taken.

For the evaluations the observation period was always taken to be composed of 120 months.

Figure 4 shows predictions of the net production for subsequent periods of six months, starting June 1974 (here June and December are the months, where predictions are evaluated). In figure 5 the subsequent predictions are evaluated in the months March and September.

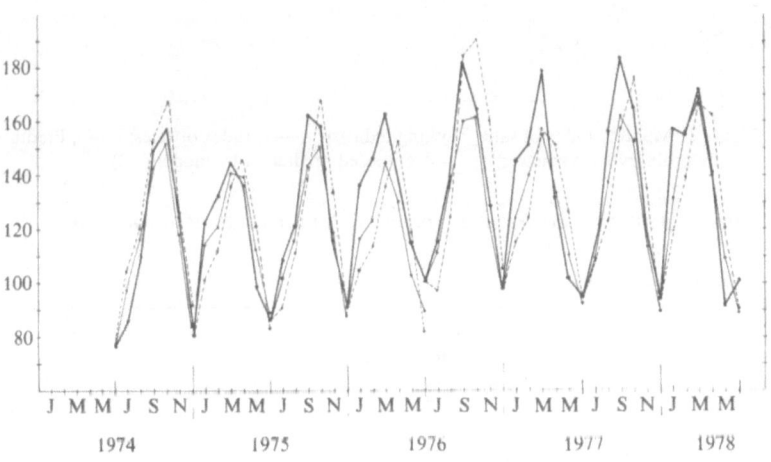

Figure 6: Monthly index of sales, clothing industry; ———, Index of sales; - - - -, Prediction by ordinary least squares; ———, Composed prediction by model (5.2).

The two figures in most cases show a considerable improvement of the proposed predictions as compared to ordinary least squares. This situation is also revealed by table 1, where the coefficients (5.1) are tabulated for each step of the calculation procedure. The first column contains the coefficients for the ordinary least squares approach, the third, fifth (and so on) column contains the generalized least squares estimations of the corresponding step.

The figures show improvements for each of the 16 predictions and from step to step. Within these examples the procedure converged very quickly. The last column of table 1 contains the inequality coef-

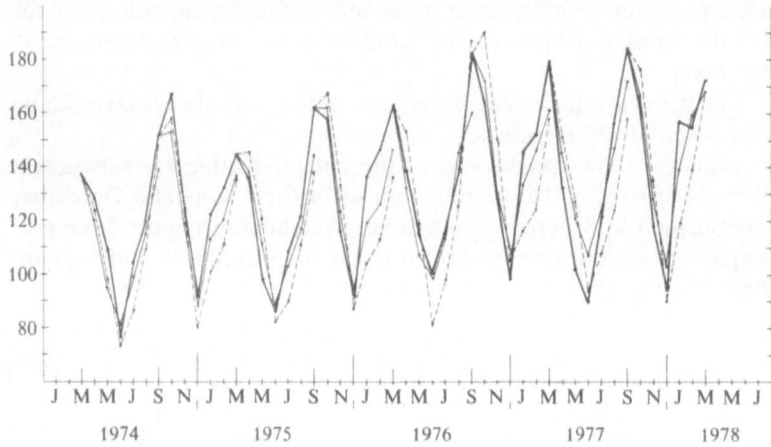

Figure 7: Monthly index of sales, clothing industry; ———, Index of sales; – – – –, Prediction by ordinary least squares; ———, Composed prediction by model (5.2).

Theil's inequality coefficients for prediction periods of six months

Table 1: Net production

| No. of last obs. | Best linear unbiased prediction Iteration step | | | | | | | | best linear prediction |
| | 1. step | | 2. step | | 3. step | | 4. step | | |
	R^1	C^2	R	C	R	C	R	C	
147	.60	.37	.52	.29	.50	.27			.47
150	.35	.21	.28	.16	.27	.15			.27
153	.57	.19	.43	.15	.42	.15			.36
156	.53	.39	.44	.33	.43	.32			.39
159	.57	.42	.47	.34					.65
162	.47	.32	.38	.26	.37	.26			.58
165	.48	.30	.35	.20	.34	.20			.78
168	.35	.21	.26	.18	.25	.17			.42
171	.36	.25	.26	.17	.24	.15			.35
174	.59	.44	.45	.35	.42	.33			.51
177	.54	.35	.42	.33	.41	.32			.29
180	.94	.28	.43	.28	.42	.28			.26
183	.35	.19	.30	.22	.30	.21			.17
186	.42	.18	.33	.17	.32	.16			.31
189	.50	.38	.43	.40	.42	.40			.24
192	.62	.52	.60	.53					.50

[1] Prediction by the regression part only.
[2] Composed prediction.

ficients for the best linear (biased) prediction according to formula (2.11), where β, Σ and σ_s are replaced by the corresponding estimators. With only a few exceptions the results are poorer than the evaluations for the best linear unbiased predictor.

Theil's inequality coefficients for prediction periods of six months

Table 2: Sales

No. of last obs.	Best linear unbiased prediction										best linear prediction
					Iteration step						
	1. step		2. step		3. step		4. step		5. step		
	R^1	C^2	R	C	R	C	R	C	R	C	
147	.31	.29	.39	.26	.46	.23	.52	.23	.58	.25	.89
150	.18	.15	.22	.15	.27	.14	.31	.14	.35	.15	.54
153	.46	.37	.52	.33	.60	.31	.67	.30	.71	.30	1.08
156	.46	.42	.47	.37	.47	.32	.48	.28	.50	.25	.82
159	.46	.42	.45	.35	.48	.30	.51	.27	.52	.26	.70
162	.32	.24	.32	.21	.36	.21	.39	.21	.40	.20	.61
165	.63	.51	.63	.45	.67	.40	.70	.37	.72	.36	.98
168	.51	.38	.51	.34	.57	.35	.62	.35	.63	.35	.87
171	.33	.27	.37	.13	.56	.13	.69	.21	.74	.24	1.05
174	.36	.26	.32	.20	.42	.14	.56	.18	.60	.20	.91
177	.51	.38	.49	.34	.54	.29	.60	.25	.63	.24	1.01
180	.50	.39	.50	.37	.55	.33	.59	.31	.60	.31	.97
183	.38	.27	.37	.23	.40	.22	.44	.23	.46	.23	.82
186	.34	.18	.35	.15	.40	.18	.45	.22	.48	.23	.77
189	.39	.33	.42	.30	.46	.26	.50	.23	.53	.21	.97
192	.45	.32	.45	.31	.48	.29	.51	.28	.53	.27	.95

[1] Prediction by the regression part only.
[2] Composed prediction.

The situation is by far not so clear-cut when we look at the sales series. Here almost always the ordinary least squares estimators deliver better performances than the subsequent generalized least squares steps. But anyway we get improved results by the composed prediction (see figure 6 and 7 and table 2). The sales series obviously needs an adequate transformation before the model (5.2) is applied. (For instance we got improved results simply by shortening the length of the observation period – we do not list the figures here.) Here the best linear predictions are very poor, as may be seen in the last column of table 2. From this fact we can conclude that obviously the best linear (biased) predictor reacts very sensitively to an unadequate model fit, whereas the best linear "unbiased" predictor seems to be more robust in this sense.

6. References

[1] Anderson, T. W. (1971), *The statistical analysis of time-series.* J. Wiley.
[2] Anderson, T. W. (1975), Maximum likelihood estimation of parameters of autoregressive processes with moving average residuals and other covariance matrices with linear structure. *Annals of Statistics* 3, 1283–1304.
[3] Box, G. E. P. and Jenkins, G. M. (1970), *Time series analysis, forecasting and control.* Holden Day.
[4] Eicker, F. (1967), Limit theorems for regressions with unequal and dependent errors. *Proceedings of the Fifth Berkeley Symposium on Mathematical Statistics and Probability* (Berkeley Calif. 1965/66). Vol. 1, 59–88. Berkeley Univ. California Press.
[5] Goldberger, A. S. (1962), Best linear unbiased prediction in the generalized regression model, *Journal of the American Statistical Association* 57, 369–375.
[6] Granger, C. W. J. and Newbold, P. (1977), *Forecasting economic time series.* Academic Press.
[7] Hannan, E. J. (1970), *Multiple time series.* J. Wiley.
[8] Hannan, E. J. and Nicholls, D. F. (1972), The estimation of mixed regression, autoregression, moving average and distributed lag models. *Econometrica* 40, 529–548.
[9] Hatanaka, M. (1974), An efficient two-step estimator for the dynamic adjustment model with autoregressive errors. *Journal of Econometrics* 2, 199–220.
[10] Hatanaka, M. (1976), Several efficient two-step estimators for the dynamic simultaneous equations model with autoregressive disturbances. *Journal of Econometrics* 4, 189–204.
[11] Heiler, S. (1971), *Wirtschaftsprognosen auf der Grundlage der Theorie schwach stationärer Prozesse.* A. Hain (Meisenheim).
[12] Heiler, S. (1973), Ökonomische Anwendungen Wienerscher Extrapolationsfunktionen. *Jahrbücher für Nationalökonomie und Statistik* 188, 152–164.
[13] Liviatan, N. (1963), Consistent estimation of distributed lags. *International Economic Review* 4, 44–52.
[14] Läuter, H. (1970), Optimale Vorhersage und Schätzung in regulären und singulären Regressionsmodellen. *Mathematische Operationsforschung und Statistik* 1, 229–243.
[15] Läuter, H. (1971), Optimale Vorhersage und Schätzung in regulären und singulären Regressionsmodellen. *Mathematische Operationsforschung und Statistik* 2, Heft 1, 69–85, Heft 2, 147–166.
[16] Läuter, H. (1971), Optimale Schätzung bei stochastischen Prozessen mit linearem Regressionsanteil. *Mathematische Operationsforschung und Statistik* 2, Heft 3, 237–246.
[17] Loeff, S. S. Van der and Leclercq, L. (1976), A note on Goldberger's best linear unbiased predictor in the generalized regression model. *La Rev. Belge de Statist. d'Inform. et de Rech. Oper.* 3, 37–42.

[18] Neftci, S. N. (1979), Lead-lag relations, exogeneity and prediction of economic time series. *Econometrica* 47, 101–113.

[19] Reinsel, G. (1979), Maximum likelihood estimation of stochastic linear difference equations with autoregressive moving average errors. *Econometrica* 47, 129–151.

[20] Schönfeld, P. (1969), *Methoden der Ökonometrie,* Band I. Franz Vahlen.

[21] Wallis, K. F. (1967), Lagged dependent variables and serially correlated errors: A reappraisal of three-pass least squares. *Review of Economics and Statistics* 49, 555–567.

[22] Willers, R. (1978), *Schwache Konsistenz von Kleinst-Quadrate-Schätzern für Regressions- und Streuungsparameter in linearen Modellen.* (Dissertation Abt. Statistik d. Univ. Dortmund.)

A Comparative Study on the Performance of Two Forecasting Techniques Based Either on Distributed-Lag Models or on Pay-off Distributions

by

Hans-J. Lenz

Berlin

Zusammenfassung:
In einigen Anwendungsfällen besteht die Möglichkeit, Lagverteilungen mittels Dichteschätzer zu schätzen. Dies ist beispielsweise der Fall, wenn die individuellen Verweilzeiten bestimmter Objekte bekannt sind. Daneben besteht die Möglichkeit, diese Daten erst zu aggregieren und dann die Lagverteilung mittels eines der bekannten Distributed-Lag-Ansätze zu schätzen. Aus Simulationsexperimenten wird deutlich, daß die Unterschiede in der Prognosegenauigkeit nahezu vernachlässigbar sind, daß aber erhebliche Unterschiede im Rechen- und Speicheraufwand bestehen.

Summary:
In several applications one can estimate distributed lags using density estimation techniques. This is for instance the case if the individual lags of a specific set of objects can be sampled. Moreover, an alternative is to aggregate the sampled data and to estimate the lag distribution using one of the usual distributed lag modelling techniques. Experiments based on simulations produce some evidence that there is no significant difference in the forecasting errors but that both approaches differ mainly in the computational efforts and the storage needed.

3*

1. Introduction

It is a matter of fact that mainly in the microeconomic environments forecasting formulas may be developed using either disaggregated or aggregated data.

To give just one example one may think of modelling the relationship between sales per day and the corresponding daily totals of payments.

This may easily be done by a distributed-lag modelling. If no prior information about special features of such a relationship are available one can try a black-box approach, cf. G. E. P. Box and G. M. Jenkins [1], to a single input – single output system. Evidently, the input corresponds to the daily sales volume and the output to daily totals of payments.

Such an approach may be called a collective approach because it start's implicitly with "collecting" or aggregating individual data, i. e. adding the totals of the single invoices per day to produce just one figure for the sales of that day and adding in a similar way the individual payments stemming from the different pay-offs.

Needless to say that the use of individual or disaggregated data is reasonable too. To develop a forecasting formula in this case a random sample of the pay-off times of the single invoices within a given time interval is needed, cf. H. Ludwig [7] and H. Langen [5]. Based on those data a direct estimation of the (discrete) pay-off density function is possible. In a straight forward manner this density can be used for forecasting, cf. Langen [5].

Using the notation:

X_{jt} total of the *j*-th invoice issued on day *t*
X_t sales per day *t*
Y_{it} value of the *i*-th payment received on day *t*
Y_t total of payments received per day *t*

the aggregation problem may be represented by the following diagram:

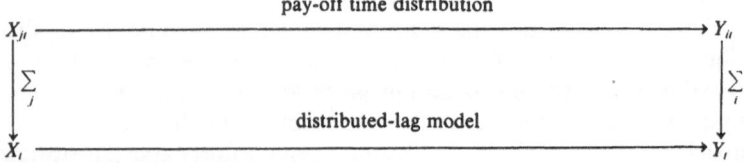

Figure 1: The aggregation problem.

As at least two competitors for forecasting the daily totals of payments now exist – the author apologizes for dropping nearly the whole set of available modern forecasting formulas – we feel that there is a challenge to analyse the behaviour and the forecasting performance of both models.

For the reader's sake we start reviewing the both approaches in some detail in chap. 2 to 3. In chap. 4 we comment on the experimental design and represent in chap. 5 the final results of some Monte-Carlo-studies.

2. The Individual Approach Using a Pay-off Time Distribution

Consider as an example a firm sending daily away the invoices to its customers and receiving daily payments from them. Our model consists mainly of the triple (N_t, X_{jt}, D_{jst}) of random variables and a set $A = \{A_1, \ldots, A_6\}$ of underlying assumptions A_i $(i = 1, \ldots, 6)$, cf. B. Streitberg [8]. Let us define the random variable N_t as the number of issued invoices per day t and the random variable X_{jt} as the total of the invoices j issued in t with $j = 1, 2, \ldots, N_t$. Both variables define the sales X_t per day t as follows:

$$(1) \qquad X_t = \sum_{j=1}^{N_t} X_{jt} \, .$$

For auditing or monitoring purposes we introduce an expiration indicator D_{jtt} defined by

$$(2) \qquad D_{j\tau t} = \begin{cases} 1 & \text{if invoice } j \text{ is issued in period } \tau \\ & \text{and is paid in period } t \geq \tau \\ 0 & \text{else} . \end{cases}$$

We make the following assumptions:

A1: *Error-free payments and no loss of debitors*

$$(3) \qquad \sum_{t=\tau}^{\infty} D_{j\tau t} = 1 \, .$$

A2: *Independence* of the pay-off time, the number of issued invoices and the total of the individual invoices.

$$(4) \qquad f_{D,N,X} = f_D \cdot f_N \cdot f_X \, ,$$

where f means probability density function, respectively.

A3: *Homogeneity* of the pay-off probabilities.

Let $p_{j\tau t}$ be the probability that the invoice j printed in τ is paid in period t. Then

(5) $p_{j\tau t} = p_{\tau t}$ for all $j = 1, 2, \ldots, N_\tau$.

A4: *Stationarity* of the pay-off probabilities

(6) $p_{\tau t} = p_{t-\tau}$ for all $\tau, t = 1, 2, \ldots$.

A5: *No discounts,* no hire-purchases etc.

Let us define the random variable Y_t by

(7) $$Y_t = \sum_{\tau=-\infty}^{t} \sum_{j=1}^{N_\tau} D_{j\tau t} X_{j\tau} .$$

The conditional expectator of Y_t in period t given the realisations of N_τ and $X_{j\tau}$ for all $\tau \le t$ and all $j = 1, 2, \ldots, N_\tau$ is defined by

(8) $$E_t(Y_t) = E(Y_t | (N_\tau)_{\tau \le t}, \ (X_{j\tau})_{j=1,\ldots,N_\tau}, \ \tau \le t) .$$

Substitution (7) into (8) and expanding the conditional expectator of Y_t gives, cf. B. Streitberg [8]

(9) $$E_t(Y_t) = \sum_{\tau=-\infty}^{t} p_{t-\tau} X_\tau = \sum_{\tau=0}^{\infty} p_\tau X_{t-\tau} ,$$

where we have used the equation

$$E_t(D_{j\tau t}) = p_{t-\tau}$$

implied by our assumption (A2).

H. Langen [5] and his followers arrived at the same linear model (9) but arguing either from a pure deterministic point of view or deducing the probabilistic set-up rather erratically, cf. R. Edin [3].

As the pay-off probabilities will be unknown in practical cases they must be estimated based on a sample of observations from $(N_t, X_{jt}, D_{j\tau t})_{t=1,\ldots,T}$. Under the above assumptions and a finite horizon $k_0 < \infty$ in (9) one may follow R. Edin's [3] approach who derived the maximum-likelihood-estimators under the hypothesis of a multinomial distribution of sales units over the different pay-off times $k = 0, 1, \ldots, k_0$:

(10) $$\hat{p}_k = \frac{\displaystyle\sum_{l=1}^{T} \sum_{j=1}^{N_l} D_{j(l-k)l} X_{jl}}{\displaystyle\sum_{l=1}^{T} X_l}$$ for all $k = 0, 1, \ldots, k_0$.

As at least two competitors for forecasting the daily totals of payments now exist – the author apologizes for dropping nearly the whole set of available modern forecasting formulas – we feel that there is a challenge to analyse the behaviour and the forecasting performance of both models.

For the reader's sake we start reviewing the both approaches in some detail in chap. 2 to 3. In chap. 4 we comment on the experimental design and represent in chap. 5 the final results of some Monte-Carlo-studies.

2. The Individual Approach Using a Pay-off Time Distribution

Consider as an example a firm sending daily away the invoices to its customers and receiving daily payments from them. Our model consists mainly of the triple (N_t, X_{jt}, D_{jst}) of random variables and a set $A = \{A_1, \ldots, A_6\}$ of underlying assumptions A_i ($i = 1, \ldots, 6$), cf. B. Streitberg [8]. Let us define the random variable N_t as the number of issued invoices per day t and the random variable X_{jt} as the total of the invoices j issued in t with $j = 1, 2, \ldots, N_t$. Both variables define the sales X_t per day t as follows:

$$(1) \qquad X_t = \sum_{j=1}^{N_t} X_{jt}.$$

For auditing or monitoring purposes we introduce an expiration indicator D_{jrt} defined by

$$(2) \qquad D_{jrt} = \begin{cases} 1 & \text{if invoice } j \text{ is issued in period } \tau \\ & \text{and is paid in period } t \geq \tau \\ 0 & \text{else}. \end{cases}$$

We make the following assumptions:

A1: *Error-free payments and no loss of debitors*

$$(3) \qquad \sum_{t=\tau}^{\infty} D_{jrt} = 1.$$

A2: *Independence* of the pay-off time, the number of issued invoices and the total of the individual invoices.

$$(4) \qquad f_{D,N,X} = f_D \cdot f_N \cdot f_X,$$

where f means probability density function, respectively.

A3: *Homogeneity* of the pay-off probabilities.

Let $p_{j\tau t}$ be the probability that the invoice j printed in τ is paid in period t. Then

(5) $\qquad\qquad p_{j\tau t} = p_{\tau t}$ for all $j = 1, 2, \ldots, N_\tau$.

A4: *Stationarity* of the pay-off probabilities

(6) $\qquad\qquad p_{\tau t} = p_{t - \tau}$ for all $\tau, t = 1, 2, \ldots$.

A5: *No discounts,* no hire-purchases etc.

Let us define the random variable Y_t by

(7) $\qquad\qquad Y_t = \sum\limits_{\tau = -\infty}^{t} \sum\limits_{j=1}^{N_x} D_{j\tau t} X_{j\tau}$.

The conditional expectator of Y_t in period t given the realisations of N_τ and $X_{j\tau}$ for all $\tau \le t$ and all $j = 1, 2, \ldots, N_\tau$ is defined by

(8) $\qquad\qquad E_t(Y_t) = E(Y_t | (N_\tau)_{\tau \le t}, (X_{j\tau})_{j=1, \ldots, N_\tau}, \tau \le t)$.

Substitution (7) into (8) and expanding the conditional expectator of Y_t gives, cf. B. Streitberg [8]

(9) $\qquad\qquad E_t(Y_t) = \sum\limits_{\tau = -\infty}^{t} p_{t-\tau} X_\tau = \sum\limits_{\tau = 0}^{\infty} p_\tau X_{t-\tau}$,

where we have used the equation

$$E_t(D_{j\tau t}) = p_{t-\tau}$$

implied by our assumption (A2).

H. Langen [5] and his followers arrived at the same linear model (9) but arguing either from a pure deterministic point of view or deducing the probabilistic set-up rather erratically, cf. R. Edin [3].

As the pay-off probabilities will be unknown in practical cases they must be estimated based on a sample of observations from $(N_l, X_{jl}, D_{j\tau t})_{l=1, \ldots, T}$. Under the above assumptions and a finite horizon $k_0 < \infty$ in (9) one may follow R. Edin's [3] approach who derived the maximum-likelihood-estimators under the hypothesis of a multinomial distribution of sales units over the different pay-off times $k = 0, 1, \ldots, k_0$:

(10) $\qquad\qquad \hat{p}_k = \dfrac{\sum\limits_{l=1}^{T} \sum\limits_{j=1}^{N_l} D_{j(l-k)l} X_{jl}}{\sum\limits_{l=1}^{T} X_l}$ for all $k = 0, 1, \ldots, k_0$.

The nominator is simply the sum over the totals of those invoices issued in the sample interval $[1, T]$ which were paid after a pay-off time of k periods. The denominator corresponds the total sales volume in $[1, T]$.

3. The Collective Approach Using a Distributed-Lag Model

Using aggregated data only, i.e. observations from the bivariate stochastic process (X_t, Y_t), where X_t and Y_t are measured as deviations from their corresponding expected values, one can derive a distributed-lag model as follows.

The total Y_t of payments received per day t is surely predetermined by the sales volumes $X_t, X_{t-1}, \ldots$ A reasonable approximation to the underlying, but unknown mechanism is to make use of a linear, parametric model with time-invariant coefficients:

$$(11) \qquad Y_t = b_0 X_{t-b} - b_1 X_{t-b-1} - \cdots - b_s X_{t-b-s} + w_t ,$$

where b is a non-negative integer parameter representing a pure time delay, s is a non-negative integer parameter corresponding the maximum lag in the model, b_i $(i = 1, 2, \ldots, s)$ are the lag coefficients and $(w_t)_{t=1, 2, \ldots}$ is a possibly auto-correlated noise process reflecting the influence of some hidden factors possibly interacting with an approximation error.

A convenient way of modelling the noise process (w_t) is to use the class of autoregressive integrated moving-average processes, cf. G. E. P. Box and G. M. Jenkins [1] defined by

$$(12) \qquad w_t = D_q(L) C_p^{-1}(L) v_t ,$$

where

$$D_q(L) = 1 - d_1 L - \cdots - d_q L^q ,$$
$$C_p(L) = 1 - c_1 L - \cdots - c_p L^p ,$$
$$L^0 v_t = v_t , \qquad L^m v_t = L^{m-1} v_{t-1}$$

with $E(v_t) = 0, \operatorname{var}(v_t) = \sigma_v^2$ and $(v_t)_{t=1, 2, \ldots}$ not auto-correlated.

Although not arguing on sound economic grounds one can introduce in (11) one part of a special eo-ipso prediction formula hoping to improve the forecasting abilities of the augmented model. For instance, make Y_t in (11) not only dependent upon the history of the sales $X_t, X_{t-1}, \ldots$ but also dependent upon its own history Y_{t-1},

$Y_{t-2}, \ldots$. Then (11) is generalized to a rational distributed-lag model

(13) $Y_t - a_1 Y_{t-1} - \cdots - a_r Y_{t-r}$

$$= b_0 X_{t-b} - \cdots - b_s X_{t-b-s} + w_t .$$

Defining the operators

$$A_r(L) = 1 - a_1 L - \cdots - a_r L^r \quad \text{and}$$
$$B_s(L) = b_0 - b_1 L - \cdots - b_s L^s$$

the whole model may be written more compactly as

(14) $A_r(L) Y_t = B_s(L) X_{t-b} + w_t$

(15) $C_p(L) w_t = D_q(L) v_t .$

Keeping in mind the equation (12), specifying the polynomials of the noise process $(w_t)_{t=1, 2, \ldots}$ as $p = q = 0$ for the sake of simplicity of notation and expanding – if possible – $A_r^{-1}(L) \cdot B_s(L)$ into the infinite sum $\sum_{\tau=0}^{\infty} g_\tau L^\tau$ the conditional expectator of Y_t in period $t-1$

$$E_{t-1}(Y_t) = E(Y_t | X_{t-1}, X_{t-2}, \ldots; Y_{t-1}, Y_{t-2}, \ldots)$$

is equal to

(16) $E_{t-1}(Y_t) = \sum_{\tau=0}^{\infty} g_\tau \hat{X}_{t-\tau-b}$

where

$$\hat{X}_{t-\tau-b} = \begin{cases} X_{t-\tau-b} & \text{if } \tau + b \geq 1 \\ E_{t-1}(X_t) & \text{if } \tau + b = 0 , \end{cases}$$

which has a striking similarity to the formula (9). Needless to say that one should be aware of the different nature of stochastic processes involved.

In the order to obtain an identifiable distributed-lag model we have to make some further assumptions, cf. M. Deistler [2], aside from those about the nature of the superimposed noise. Firstly, all the polynomials involved, i.e. A, B, C and D, set equal to zero must have roots outside the unit circle and neither the polynomials A, B nor the polynomials C, D are allowed have any common roots. Secondly, the input process $(X_t)_{t=1, 2, \ldots}$ must be persistent existing in Åström's sense, cf. M. Deistler [2], in order to produce "sufficient" deviations in the system and any cross-correlation between the input process

(X_t) and the error process (v_t) is excluded. Moreover, making the unrealistic assumption that the orders of the polynomials and the delay factor b is known all the parameters involved may be estimated uniquely given a sample of observations of (X_t, Y_t) and using for example a proxy to the conditional likelihood estimators under the hypothesis of a normally distributed error process, cf. G. E. P. Box and G. M. Jenkins [1]. In practical applications one has to specify the orders of the polynomials r, s, p, q and the delay factor b. There are now several methods available for selecting the orders of the polynomials, using only the information from the data, cf. H.-J. Lenz [6].

4. The Experimental Design

It seems impossible to design a bivariate time-series generator for our purpose which is unbiased in the sense that it doesn't favour one of the approaches mentioned above. After weighing the pros and cons of some few methods we decided to produce the needed data by generating the disaggregated data first of all assuming some specific distributions for the pay-off time, the number of invoices per day and the total amount per invoice. The aggregated data are then devided simply by aggregating the generated totals per invoice and sums of payment within each single day during the simulation period. A global lay-out of the used generator is represented in the following diagram.

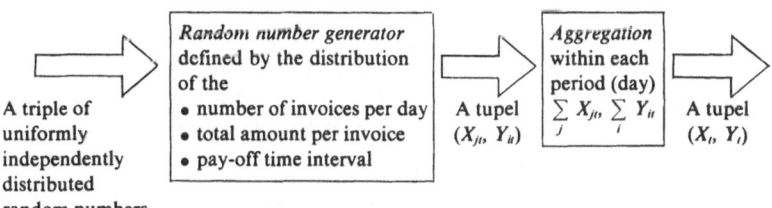

Figure 2: Global lay-out of the used random number generator.

To reduce the complexity of the experimental design the following factors were fixed to one specific level only:

(1) The random variables number of invoices, total amount per invoice and pay-off time are independently distributed.

(2) The random variable number of invoices is uniformly distributed over the set $\{0, 1, 2, 3, 4, 5\}$.

(3) The random variable total amount per invoice is distributed according to the following distribution (Table 1).

Table 1: The distribution of the total amount per invoice

classes	relative frequencies $\times 10^{-3}$
[0, 5]	0
(5, 10]	100
(10, 20]	200
(20, 30]	200
(30, 40]	250
(40, 50]	125
(50, 75]	75
(75, 100]	50

One may easily compute the median $\hat{\mu} = 30$, the mean value $\mu \approx 32$ and a standard deviation $\sigma \approx 19$ and detects a slightly skewed distribution. The combination (2) and (3) is chosen in such a way that the compound distribution of the resulting sales volume is characterized by a modest size of the standard deviation.

(4) By the error and trial method the length of the generated time series is fixed to 100 (200) data points being a compromise between the large sample effect on the precision of specification and estimation and the computing times.

Main attention was paid to the influence of the type of distribution of the pay-off time spans. The type of distribution is characterized by its expected value, its standard deviation and the skewness. The following levels of the three factors were actually analyzed (Table 2).

In all the cases the delay time factor b was set equal to $b = 3$ except for the distribution IIc with $b = 4$.

As criteria for the comparison one may choose the standard deviation of the one-step-ahead forecasting error, i.e.

$$(17) \qquad SE = \sqrt{\frac{1}{N} \sum_{t=1}^{N} (a_t - \bar{a})^2}$$

Table 2: Type of distributions of the pay-off time spans used for simulation

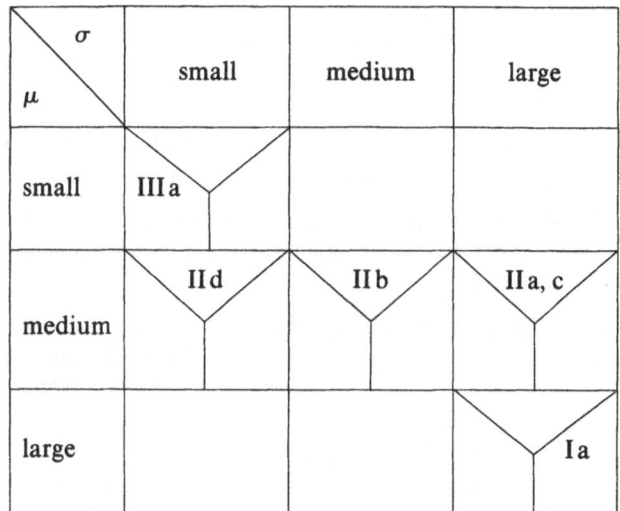

	small	medium	large
small	III a		
medium	II d	II b	II a, c
large			I a

> I a: $\mu = 6.7$, $\sigma = 0.85$, skewed to the right
> II a: $\mu = 6.0$, $\sigma = 0.81$, symmetric
> II b: $\mu = 6.0$, $\sigma = 0.73$, symmetric
> II c: $\mu = 6.0$, $\sigma = 0.94$, symmetric
> II d: $\mu = 6.0$, $\sigma = 0.59$, symmetric
> III a: $\mu = 5.3$, $\sigma = 0.57$, skewed to the left

with $\bar{a} = \dfrac{1}{N} \sum\limits_{t=1}^{N} a_t$ and $a_t = y_t - \hat{E}_{t-1}(Y_t)$ and N is equal to the number of data points generated.

As the approach via the pay-off time distribution will always produce non-negative weights (probabilities) $\hat{p}_k$ summing up to $\sum\limits_{k=0}^{k_0} \hat{p}_k = 1$, it makes sense to estimate the steady state gain $\hat{R} = \sum\limits_{\tau=0}^{\infty} \hat{g}_\tau$ when using the distributed-lag model approach.

From the economic theory point of view the specification and estimation of distributed-lag models should be only feasible if the side constraint $\hat{R} = 1$ and all the weights $\hat{g}_\tau$ for $\tau = 0, 1, 2, \ldots$ are non-negative. The first condition means that there is no loss of payments, i.e. the sales and the payments are balanced in the long run, cf. (A1). The assumption of non-negative weights is self-evident because an increase in the sales volumes should always cause ceteris paribus a marginal increase in the volume of payments. Nevertheless we relax these constraints mainly because of

(a) the difficulties which are caused by a constrained specification and estimation procedure if using distributed-lag models and

(b) the fact that our main objective is to find out the forecasting abilities of both approaches rather than to give a proper explanation of the underlying mechanism.

Because of the fact that the approach to design a forecasting formula using the estimated pay-off time span distribution automatically delivers non-negative and normalized weighing coefficients $\hat{g}_\tau$ whereas the other approach relaxes these constraints it seems plausible that the solutions of a constrained problem should not be better than those of an unconstrained problem. But again, the problems or approaches are not equivalent in a proper sense as pointed out in chap. 1.

5. Final Results

Table 3 summarizes some representive results of the simulation studies, cf. Th. Hübner [4].

Our main conclusions are:

(1) The differences in the standard deviations (SE) are nearly negligeable although there is some evidence (Figure 3) for a weak dominance of the distributed-lag model approach.

(2) Almost everywhere are the orders p, q of the noise process equal to zero, i.e. the design condition of a non-autocorrelated noise superimposed on the input X_t is rediscovered.

(3) The delay factors $\hat{b}$ are positively biased compared to $b=3$ except for series of type IIc where $b=4$.

(4) In most of the cases there is no need for denominator dynamics, i.e. the order r of the polynomial $A_r(L)$ can be set equal to zero as may be expected by an economic reasoning.

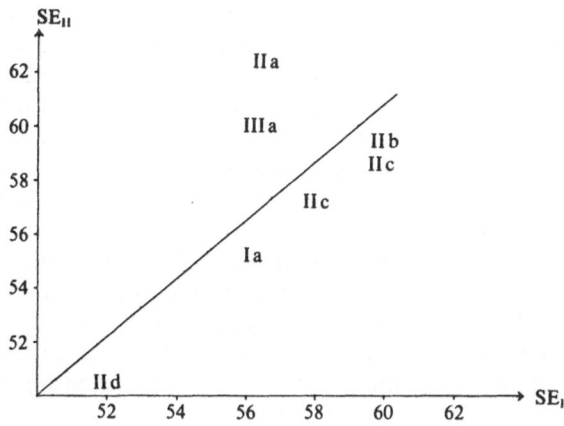

Figure 3: A scatter diagram of the standard deviation in error of both methods.

Table 3: Main results of the simulation study

Type of distribution	Pay-off time distribution (I)		Forecasting method Distributed-lag model (II)						
	SE_I	$\hat{R}_I$	$\hat{r}$	$\hat{b}$	$\hat{s}$	$\hat{p}$	$\hat{q}$	SE_{II}	$\hat{R}_{II}$
I a	56	1	0	7	1	0	0	55	0.77
II a	56	1	0	6	1	0	0	62	0.75
II b	60	1	0	3	4	3	0	58	0.95
II c	58	1	0	6	1	0	0	57	0.76
II d	52	1	0	3	5	0	0	51	1.01
II c	60	1	0	5	1	0	0	59	0.78
III a	57	1	1	4	1	0	0	61	0.47

(5) The estimated gain factor $\hat{R}$ is almost negatively biased, i.e. less than one. This is the response to relax the side constraints of the non-negative weights and a normalized sum of the weights (coefficients of the lag distribution).

78   *H.-J. Lenz*

Trying to resume the main message of the study the main discrepancies doesn't seem to lie in a different performance of both methods as long as statistical criteria are used. The main difference is caused by the different methodologies and its different degree of practicability. The distributed-lag approach only make use of "some few" aggregated data like a bivariate time series (low need for storage facilities) while using a very sophisticated technique for developing the appropriate forecasting formula. Quite opposite the method of pay-off time distributions is rather uncomplicated and simple to apply. However, a data base seems necessary to store all the sample values used for this approach. With reference to the acceptability of a forecasting formula a potential forecaster should be advised to balance carefully the cost of computing and the cost of storing before choosing one of the prediction formulas presented above.

Acknowledgements

I thank Dipl.-Kfm. Th. Hübner very much who had carried out the computational burden with skillfullness and patience during quite a lot of never ending nights in front of a noisy computer.

6. References

[1] Box, G. E. P. and Jenkins, G. M. (1970), *Time series analysis forecasting and control*, Holden-Day, San Francisco etc., 1970.
[2] Deistler, M. (1975), z-transform and identification of linear econometric models with autocorrelated errors, *Metrika* 22.
[3] Edin, R. (1969), Übergangsfunktionen in betriebswirtschaftlichen Systemen, *Zeitschrift für Betriebswirtschaft* 39.
[4] Hübner, Th. (1978), *Kurzfristige Finanzplanung mit Hilfe von Box-Jenkins Techniken oder Verweilzeitverfahren? Ein Simulationsvergleich anhand von Umsatzeinnahmen*, M. A. Thesis, Free University of Berlin.
[5] Langen, H. (1964), Die Prognose von Zahlungseingängen. Die Abhängigkeit der Bareinheiten von Umsätzen und Auftragseingängen in dynamischer Betrachtung, *Zeitschrift für Betriebswirtschaft* 34.
[6] Lenz, H.-J. (1978), *Strategies of implementation of a fully automatic Box & Jenkins forecasting technique*, Discussion Paper No. 7/78, Institut für Quantitative Ökonomik und Statistik, Free University of Berlin.
[7] Ludwig, H. (1932), Die Einnahmenseite des Finanzplanes, *Zeitschrift für Handelswissenschaft und Handelspraxis*
[8] Streitberg, B. (1979), *Prognose im Langen'schen Verweilzeitmodell*, Diskussionsarbeit Nr. 7/79, Institut für Quantitative Ökonomik und Statistik, Free University of Berlin.

Updated Time Series and Econometric Forecasts

by

Bernd Schips

St. Gallen

Zusammenfassung:
Prognosen mit strukturellen Modellen können auf konventionelle Weise nicht alle die für kurzfristige Vorausschätzungen zur Verfügung stehenden Informationen adäquat nutzen. Damit sind solche Vorausschätzungsversuche prinzipiell eklektischen Vorgehensweisen unterlegen. In diesem Beitrag wird eine statistisch-methodisch befriedigende Vorgehensweise dargestellt, die durch Erweiterung struktureller ökonometrischer Modelle durch ein Beobachtungsmodell diese Nachteile ökonometrischer Prognosemodelle kompensiert. Am Beispiel eines kleinen ökonometrischen Modells wird ein Vorschlag für eine praxisgerechte DV-technische Realisierung dieser Methode gemacht, so daß von dieser Seite her alle Voraussetzungen für ernsthafte Bemühungen um eine Verbesserung der kurzfristigen Vorausschätzungen der gesamtwirtschaftlichen Entwicklung gegeben sind. Es fehlt jedoch noch eine umfassende Erfahrungspraxis unter realistischen Anwendungsbedingungen und Zielsetzungen.

Summary:
Predictions based on structural models cannot adequately consider all information available for short-term predictions. Therefore, this kind of forecasting is inferior to eclectical procedures. This paper describes a procedure that is satisfactory from the statistical point of view: the disadvantages of econometric forecasting are compensated by extending structural models by an observation-model. The example of a small econometric model shall be a proposition for a practicable EDP-technical realisation of this method. From this side, all conditions are set for serious efforts to improve short-term predictions of the economic development. Still missing is, however, an extensive experience under realistic conditions of application and aims.

In economic forecasting of the official and semi-official institutions of the Federal Republic of Germany econometric models do not play the role they should and could play. In many other Western European countries the situation is quite similar[1]. The situation in USA, however, is somewhat different[2].

It certainly would lead too far to discuss here the various reasons for this unsatisfactory situation. It may be a proper point of departure, however, to start with one of major arguments used against the use of interdependent econometric models for describing economic developments: the predicted results are obviously bad in comparison to the results obtained from time series approaches and above all, are bad in comparison to the so called eclectical procedures which rely on identities.

The comparisons of results derived from short-term predictions[3] – 'short-term' means here that the prediction period is equal or, at most, twice the 'basis-period' used in the model – usually show clear advantages when time series models are used. This is true as long as different approaches included in a comparison are drawn up to describe the development of variables represented by data with a periodicity shorter than a year.

Looking more closely at the results of time series models they loose much of their glamour as soon as all dependent variables included in more comprehensive econometric models are allowed for and not only a reduced set of some particularly important economic values is used. From an economist's point of view, the exclusive use of such a time series model is, in a way, nothing less than a capitulation. The economic model plays the second fidel vis-a-vis the stochastic model. And considering the efforts required to identify the stochastic process responsable for the development of an economic variable, we may safely assume that similar efforts devoted to econometric model building would ultimately lead to more meaningful results also in this area. Moreover, results obtained by econometric models for short-term prediction may be considerably improved if the assumption of time-invariant coefficients in structural models is given up[4].

We may also be quite explicite on one point: methods of modern time-series-analysis play an important role in the specification process of a structural econometric model, for instance, for the analysis of lag-relationships, of aspects of dynamic behaviour, etc.[5].

Though some of the fundamental advantages of structural models largely depend on the quality of the specification process we will not deal with it here explicitly. The lacunae in economic theory, the deficiencies in our data, and the stochastic nature of the problem raise fundamental difficulties when an explanation of basically economic phenomena with an econometric model is attempted. We should not overlook this fact. Often enough, a mere data analysis is predominant when econometric models are constructed. The results that appear meaningful from a statistical point of view, are subsequently 'semantically coloured' – using an expression by G. Menges [12]; but actually they are not interpreted in a genuine economic-theoretical sense. Often autoregressive models are used relying on the theory of adaptive expectations. However, as a rule, they are little more than data-transformations in order to improve the descriptive quality of the specification approaches. Further analysis of the estimated model is usually not undertaken. Policy simulations are carried out without considering that projections towards policy objectives do not have the same dynamic stability as predictions based on the usual structural model.

Our brief discussion already shows that there remains a large number of challenging issues – yet to be explored. Our following discussion will focus on but one of these aspects: the predictive quality of econometric models.

Generally we may say that the predictions based on econometric models yield optimal predictions for a large class of loss-functions[6]. For this to be true, it is necessary to have a proper specification of the model as well as a proper estimation of the coefficients adequate for the structure of the model. Though these problems are not dealt with here explicitly, they should not be ignored.

Further conditions for optimal econometric prediction are that no additional information is available and the values of the exogenous variables are known. This last point shall also be excluded here; though also here the methods of time series analysis may give quite adequate results for the problems mentioned.

Our discussion will revolve on the issue of how additional information not included in the estimated model can be used for the purpose of prediction. This additional information includes in particular
– a priori information about the development of the dependent variables
– tentative values for a part or for all variables of the model.

If this additional information exists – and in the predicting business this is normally the case (mostly in the form of monthly values and closely related to model variables with a longer basis period, for instance, with regard to updated time series as unfilled orders, indices of output, various price indices, etc.) – then major advantages of

those prediction methods become evident, which transform this information on the basis of identities, given certain assumptions about the
exogenous variables, in a successive process of iteration into more or
less usefull, plausible and consistent predictions[7].

Even when using the same values for exogenous variables in econometric models as in the eclectical procedures, a disadvantage for the
econometrician will be evident since he will not be in a position to use
the additional information in a conventional way when attempting
his predictions. Such additional information can, however, be included in the eclectical model since the respective variables are
treated as exogenous variables while in an econometric model they
are endogenous.

These basic disadvantages – and our daily experience suggests this
a priori advantage of the eclectical procedure to be real – will remain
as long as we will not be able to adequately include in econometric
models additional information about relevant variables that is not already contained in the estimated model.

The fine-tuning frequently mentioned in this context, can only in-
adequately achieve this; and it does so in an unsatisfactory way. The
techniques usually proposed for adjusting coefficients based on prediction experience do not satisfy scientific criteria. Such mechanical
devices hardly can stand the test in reality, and divert from the real
problems[8].

However, a proper solution for the inclusion of additional information exists, as will be outlined in the following.

Our starting point is a (not necessarily linear) econometric model
of the form

$$y(t) = F(y(t), z(t), \alpha) + u(t).$$

$y(t)$ is the $(N \times 1)$-vector of the dependent variables, $z(t)$ is the
$(K \times 1)$-vector of the predetermined variables, $u(t)$ is the $(N \times 1)$-vec-
tor of the disturbance variables of the model, and the parameter α
represents all unknown parameters of the functional form of the
model.

Based on the observations for $y(t)$ and $z(t)$, $t = 1, \ldots, T$, an estimation $\hat{\alpha}$ for α is made. The estimated model

$$y(t) = F(y(t), z(t), \hat{\alpha}) + \hat{u}(t)$$

includes also its reduced form which cannot necessarily be written in
explicite form,

$$y(t) = P(z(t), \hat{\alpha}) + \hat{v}(t).$$

This latter form is useful for making ex post predictions,

$$\hat{y}(t) = P(z(t), \hat{\alpha}), \quad t = 1, \ldots, T.$$

Of more interest, however, are predicted values $y(t)$ for the dependent variables $y(t)$ for $t = T+1, T+2, \ldots, T+\tau, \tau \in \mathbb{N}$ which are determined by the corresponding values for the predetermined variables by using the reduced form.

A model prediction $y(t)$, $t = T+1, T+2, \ldots, T+\tau$ is obviously based on the information provided by the model and the predetermined variables. However, whenever additional external knowledge exists about the values of some variables $y(t)$ for $t = T+1, T+2, \ldots, T+\tau_p, \tau_p \leq \tau$, this additional information should be used to correct the predicted values $\hat{y}_1^p(t), \ldots, \hat{y}_{N_p}^p(t)$, $N_p \leq N$ of the dependent variables $y_1^p(t), \ldots, y_{N_p}^p(t)$.

A procedure useful for such a correction is one based on the Kalman-filter-technique. When using this method, a distinction between a state-model and an observation-model is introduced. The non-observable variables of the state-model are estimated on the basis of the values of the observation-model which relies on stochastic control theory[9]. In econometrics, this has been especially applied for the estimation of variable coefficients[10].

For our problem, the state-model is given by

$$y^p(t) = \hat{y}^p(t) + \vartheta^p(t)$$

with

$$E(\vartheta^p(t)) = 0$$

and

$$E(\vartheta^p(t)\,\vartheta^p(t)') =: V(t)$$

for all $t = 1, \ldots, T+\tau_p$.

The observation-model describes formally the relationship between N_a observable variables $y_1^a(t), \ldots, y_{N_a}^a(t)$ that include the additional information not included in the model, and the dependent variables $y_1^p(t), \ldots, y_{N_p}^p(t)$[11], in the form

$$y^a(t) = H y^p(t) + g(z(t)) + w(t)$$

with

$$E(w(t)) = 0$$

and

$$E(w(t)w(t)') =: W(t)$$

for all $t = 1, \ldots, T+\tau_p$. H is an $(N_a \times N_p)$-matrix and g is a mapping of $\mathbb{R}^K$ in $\mathbb{R}^{N_a}$.

In a way, the econometric model is extended by a partial model in order to deal with additional information. What is left is to show how the variables $y_1^p(t), \ldots, y_{N_p}^p(t)$ can be chosen, how a observation-model can be built, how the variance-covariance-matrices $V(t)$ and $W(t)$ can be determined and how the correction-algorithm looks like. Our following discussion will focus on these problems. We will conclude our discussion by providing a little example which shows how the procedure outlined is applied in praxi.

By definition of the state-model, each variable $y_n^p(t)$, $n = 1, \ldots, N_p$ corresponds with its predicted value $\hat{y}_n^p(t)$ up to an error $\vartheta_n^p(t)$ which has an expectation value of zero. Hence, we require $y_n^p(t)$ to have an average prediction error

$$\frac{1}{T} \sum_{t=1}^{T} (y_n^p(t) - \hat{y}_n^p(t))^2$$

with a *bias*-value of zero, when an ex post prediction for the estimation period is made

$$\frac{\left[\dfrac{1}{T} \sum_{t=1}^{T} (y_n^p(t) - \hat{y}_n^p(t)) \right]^2}{\dfrac{1}{T} \sum_{t=1}^{T} (y_n^p(t) - \hat{y}_n^p(t))^2} \sim 0 \,.$$

A correction of the predicted values $\hat{y}_n^p(t)$, $t = T + \tau_p$ of the variables $y_n^p(t)$, $n = 1, \ldots, N_p$ may be made based on a priori information provided by the observations of N_a variables $y_1^a(t), \ldots, y_{N_a}^a(t)$ at $t = T + 1, \ldots, T + \tau_p$; viz. $\tau_p \leq \tau$ is ascertained by the available information for the variables $y_1^a(t), \ldots, y_{N_a}^a(t)$[12].

In order to give some idea of how to construct an observation-model, let us assume that for each variable $y_n^a(t)$, $n = 1, \ldots, N_a$ the relationship with the N_p dependent variables $y_1^p(t), \ldots, y_{N_p}^p(t)$, with the K predetermined variables $z_1(t), \ldots, z_K(t)$ and with the K' variables $x_1(t), \ldots, x_{K'}(t)$ not yet included in the original model is given by the equation

$$y_n^a(t) = [H(t)]_n y^p(t) + g_n(t) [z(t), x(t)] + w_n(t)[13] \,.$$

This equation may be further simplified. In most cases $\tau_p = 1$; thus, the inclusion of t at $H(t)$ is superfluous. If $\tau_p > 1$, we may assume that the functional relationship is constant within a given period; hence, we may start from an invariant $(N_a \times N_p)$-matrix H.

- Similar considerations apply for $g_n(t)$, and we presuppose $g_n(t) = g_n$.
- Generally speaking, we assume that

$$g_n(z(t), x(t)) = g'_n(z(t)) + g''_n(x(t))$$

is true. If so,

$$\bar{y}^a_n(t) = y^a_n(t) - g''_n(x(t))$$

can be taken instead of the original $y^a_n(t)$. This implies that on the right hand side of the equation of the observation-model only the variables of the original model will appear.

In simplified notation, we get

$$\bar{y}^a_n(t) = [H]_n y^p(t) + g_n(z(t)) + w_n(t).$$

Combining the set of equations for all $n = 1, \ldots, N_a$ variables within a system, we get the observation-model. It is important to note that all equations must be linear in $y^p_1(t), \ldots, y^p_{N_p}(t)$[14].

Having discussed the problems thus far in a more formal way, we may now give two specific examples that prove to be particularly important when equations for an observation-model used for prediction are to be found.

- If we have reasonably reliable values for $y^p_s(t)$, $s \in \{1, \ldots, N_p\}$, for $t = T+1, \ldots, T+\tau_p$, we may interpret them as observations of a variable $y^a_n(t)$. Then, we have

$$[H]_n := [0, \ldots, 0, 1, 0, \ldots, 0]$$
$$s^{\text{th}} \text{ position}$$

and

$$g_n = 0.$$

- Or, if $y^p_s(t)$ is explained by an identity which is linear in $y^p_1(t), \ldots, y^p_{N_p}(t)$ the right hand side of the equation of the observation-model corresponds with the right hand side of this identity.
- In case a variable $y^p_s(t)$, $s \in \{1, \ldots, N_p\}$ can be represented by various independent $y^p_1(t), \ldots, y^p_{N_p}(t)$ linear identities, the thus achieved conformity can be expressed; for instance, if we have a model which describes the GNP by three different identities referring to the demand, supply and distribution side than we operate, for instance, with two independent identities

$$0 = \text{GNP (supply-side)} \quad - \text{GNP (distribution-side)}$$
$$0 = \text{GNP (distribution-side)} - \text{GNP (demand-side)}.$$

The corresponding values $y_n^a(t)$ and $w(t)$ are defined by

$$y_n^a(t) = 0$$

and

$$w_n(t) = 0.$$

For all other cases, the relationships between the variables that are considered as indicators $y_r^a(t)$, $r \in \{1, \ldots, N_a\}$ and the chosen dependent variables $y_n^p(t)$, $n = 1, \ldots, N_p$ should be dealt with carefully and adequately. Such a procedure corresponds to the postulate of rationally handling and using existing information and as such relates to the present discussion in economic theory concerning the role of rational expectations in economic models.

When determining the variance-covariance-matrices $V(t)$ and $W(t)$, the following considerations shall be made. Assuming that for all $t = 1, \ldots, T, T+1, \ldots, T+\tau_p$, $V(t) = V$ is valid, it will be possible to estimate the elements v_{ij}, $i,j = 1, \ldots, N_p$ of the variance-covariance-matrix V with the aid of residuals $\hat{v}_n(t) = y_n^p(t) - \hat{y}_n^p(t)$ which are determined by ex post prediction, by

$$\hat{v}_{ij} = \frac{1}{T} \sum_{t=1}^{T} \hat{v}_i(t)\, \hat{v}_j(t).$$

Since $W(t)$ is the variance-covariance-matrix of the residuals $w(t)$ of the observation-model, we may express the 'reliability' of the observation-model with the matrix $W(t)$. We may assume that the residuals of different equations are uncorrelated; this is supported by the independency of the equations that are chosen for the observation-model.

It remains to determine the variances $s_n^2(t)$, $n = 1, \ldots, N_a$ of the various residuals $w_1(t), \ldots, w_{N_a}(t)$, whereby the special values represent the a priori information about the reliability of the relationships of observation. The following examples may highlight this:
– 'absolutely' certain

$$s_n^2(t) = 0$$

– deviations of the magnitude $\pm c$ from the observation value are possible:

$$s_n^2(t) = c^2$$

– a relative error of $p\%$ is possible:

$$s_n^2(t) = \left[y_n^a(t) \cdot \frac{P}{100} \right]^2.$$

This information for $t = T, \ldots, T + \tau_p$, must be given in order to get a diagonal matrix $W(t)$ for the respective points of time.

As for the correction algorithm, starting from the general Kalman-filter approach, with the state-model

$$s(t) = A(t)s(t) + f(t) + v(t)$$

and the observation-model

$$m(t) = H(t)s(t) + g(t) + w(t)$$

we have

$$s(t) = y^p(t)$$
$$f(t) = \hat{y}^p(t)$$
$$A(t) = 0$$
$$E(v(t)v(t)') = V(t) = V$$

and

$$m(t) = y^a(t)$$
$$g(t) = g(z(t))$$
$$H(t) = H$$
$$E(w(t)w(t)') = W(t), \text{ diagonal-matrix}$$

for all $t = T, \ldots, T + \tau_p$.

For the corrected prediction value $\hat{y}^p(T + t)$ for $y^p(T + t)$, viz. for the prediction value allowing for a priori information $y^a(T + t)$, $t = 1, \ldots, \tau_p$, we have

$$\hat{y}^p(T + t) = \hat{y}^p(T + t) + K(T + t)$$
$$\cdot [y^a(T + t) - H\hat{y}^p(T + t) - g(z(T + t))]$$

with

$$K(T + t) = VH'[HVH' + W(T + t)]^{-1}$$

for $t = 1, \ldots, \tau_p$.

It should be noted that because of $A(t) = 0$ only the values $\hat{y}^p(T + t)$ are relevant for the computation of $y^a(T + t)$; therefore, also a smoothing does not cause any change in the values for $\hat{y}^p(T + t)$[15].

This correction algorithm leads to best linear unbiased estimation functions for dependent variables predicted on the basis of an econometric model that is extended by an observation-model. Optimal prediction makes use of the predictions $\hat{y}^p(T+t)$ of the structural model as well as of the additional available information provided by the correction term $K(T+t)[y^a(T+t) - H\hat{y}^p(T+t) - g(z(T+t))]$.

The following example – admittedly with a simple structural model – will nevertheless demonstrate the possible use of the approach outlined[16]. It goes without saying that a serious application requires an econometric model that is adequately specified and that usually serves also purposes other than those of prediction as well as the formulation of an adequate observation-model. This presupposes a firm grasp of economic relationships and a fair knowledge of statistical concepts. Therefore, particularly economists well acquainted with the facts gained by applying eclectical procedures should deal with it.

Of course, applied models may soon become quite large and comprehensive. Thus, a clear representation demands such a small demonstration model. The example chosen here is a model with the name 0,0. It consists of three behavioural equations for private consumption PV, import EF, investment AIA, and an identity for the gross national product BSP[17]. The reduced form of this structural model gives the necessary state-model. The steps are as follows: preparation of the data for the estimation of coefficients, the estimation of the coefficients, an ex post prediction or a prediction simulation, the preparation of the prediction data and finally the prediction itself[18]. The results of the ex post prediction shall be called 0,3, and the prediction result 0,5.

Vis-a-vis this state-model, an observation-model is formulated. The observation-model is called 0,1. Here, it relies on the assumption that information about the future development of the two variables PVA and BSPA is directly available. For the variable PVA a respective identity is used. The value for BSPA which may be assumed to be reasonably accurate is interpreted as an observation of the variable BSP.

The two models are shown in the following. The prints conclude with a brief outline of the further steps. In doing so, the necessary ex post or respectively prediction results for the structural model as well as an assessment of the bias-part of the ex post prediction will be attempted. A specification of the degree of reliability of the external information used in the observation-model is required from the user. The available input options are provided by the protocol. The shortened version of the prediction results is juxtaposed with the original prediction made with the structural model with the predictions corrected using external information[19].

```
GIB KOMMANDOSn:nMUEB.,MODELL.1n.


START MODELL (3.29) n:0,On.

GIB NUMMERN DER AUSGABEMEDIENn:9,9n.
 1 ***  - DEMONSTRATIONSMODELL FUER DAS MEBA-SYSTEM: BO-MODELL I
 2 ***  -
 3 ***  -
 4  1        PV=LABS(VED,LAG(PV,1),D1);
 5  2        EF=LABS(BSP,TOT,D1);
 6  3        AIA=LABS(PV,LAG(AEU,1),LAG(AIA,1),D1);
 7  4        BSP:=PV+AIA-EF+Z4;

UEBERSETZUNG FEHLERFREI
STOP

ENDE MODELL (3.29) 0.76

GIB KOMMANDOSn:nMUEB.,MODELL2n.


START MODELL (3.29) n:0,1n.

GIB NUMMERN DER AUSGABEMEDIENn:9,9n.
 1  1  PVA:=BSP-AIA+EF-Z4;
 2  2  BSPA:=BSP;

UEBERSETZUNG FEHLERFREI
STOP

ENDE MODELL (3.29) 0.57

GIB KOMMANDOSn:=nSTARTE,DATEI=*1000n.

START STDHP

GIB NAMEN DES ZUSTANDSMODELLS - Z,2n:0,On.

GIB FUER DAS ZUSTANDSMODELL FOLGENDE ERGEBNISSE:
EXPOST-PROGNOSE-ERGEBNIS
PROGNOSEERGEBNIS  *** ZEILENWEISE ****
n:0,3
0,5n.

GIB NAMEN DES BEOBACHTUNGSMODELLS - Z,2n:0,1n.
SOLLEN WEITERE ZU KORRIGIERENDE GEM.ABH. VARIABLEN AUS DEM ZUST.MOD.
HINZUGEFUEGT WERDEN - (J/N)n:Nn.

     VARIABLE         BIAS-ANTEIL

        BSP            0.0

        AIA            0.0

        EF             0.0

PROGNOSEANFANG      : 7401
MAX.ANZ.D.PROGNOSEN :    3
GIB ANZ.D.PROGNOSENn:2n.

GIB FUER DIE FOLGENDEN VARIABLEN   2 WERTE --- BEGINNEND 7401
SODANN  FUER  VARIANZ ABSOLUT    - 1
        FUER ABWEICHUNG ABSOLUT  - 2
        ODER ABWEICHUNG PROZENTUAL - 3
SCHLIESSLICH DIE ZUGEHOERIGEN WERTE
JEDE EINGABE MIT FLS ABSCHLIESSEN
     PVAn:337.78,346.48n.n:2n.n:1,0.5n.
     BSPAn:595.6,576.6n.n:2n.n:0.5,1n.
```

```
*** PROGNOSEKORREKTUR ***
    ZUM ERGEBNIS:  0  5


         VARIABLE:        BSP

         URSPR.          KORR.
         PROGN.          PROGN.

7401     613.173         597.144
7501     600.397         583.484


         VARIABLE:        AIA

         URSPR.          KORR.
         PROGN.          PROGN.

7401     79.946          76.103
7501     81.143          77.242


         VARIABLE:        EF

         URSPR.          KORR.
         PROGN.          PROGN.

7401     186.947         194.118
7501     184.013         192.411
STOP

ENDE STDHP   1.80
```

It follows that already little external information which cannot be included in structural econometric models but always is available in some form or other, may lead to an important inprovement of the short term forecasting results. Another special case is the exclusive consideration of the residuals, i.e. of the prediction errors encountered in ex post prediction or prediction simulation. This is shown by way of an example with the variable PV. The observation-model is, for this purpose, extended by this variable. The procedure may be seen directly from the dialogue about the applied EDP-System[20].

```
GIB KOMMANDOSɑ:                    ɑSTARTE,DATEI-*1000ɑ.

START STDHP

GIB NAMEN DES ZUSTANDSMODELLS - Z,Zɑ:0,0ɑ.

GIB FUER DAS ZUSTANDSMODELL FOLGENDE ERGEBNISSE:
EXPOST-PROGNOSE-ERGEBNIS
PROGNOSEERGEBNIS  ***  ZEILENWEISE  ****
ɑ:0,3
0,5ɑ.

GIB NAMEN DES BEOBACHTUNGSMODELLS - Z,Zɑ:0,1ɑ.
SOLLEN WEITERE ZU KORRIGIERENDE GEM.ABH. VARIABLEN AUS DEM ZUST.MOD.
HINZUGEFUEGT WERDEN - (J/N)ɑ:Jɑ.
GIB DIE VARIABLEN EINɑ:AUS,AIA,PV;ɑ.
DIE   1-TE VARIABLE IST NICHT GEM.ABH. IM ZUST.MOD.
SIE WURDE GESTRICHEN
DIE   2-TE VARIABLE IST BEREITS IM BEOB.MOD.
SIE WURDE GESTRICHEN


    VARIABLE            BIAS-ANTEIL

       BSP               0.0

       AIA               0.0

        EF               0.0

        PV               0.0

PROGNOSEANFANG      : 7401
MAX.ANZ.D.PROGNOSEN :    3
GIB ANZ.D.PROGNOSENɑ:2ɑ.

GIB FUER DIE FOLGENDEN VARIABLEN    2 WERTE --- BEGINNEND 7401
SODANN  FUER  VARIANZ ABSOLUT       - 1
        FUER ABWEICHUNG ABSOLUT     - 2
        ODER ABWEICHUNG PROZENTUAL  - 3
SCHLIESSLICH DIE ZUGEHOERIGEN WERTE
JEDE EINGABE MIT FLS ABSCHLIESSEN
      PVAɑ:337.78,346.48ɑ.ɑ:3ɑ.ɑ:0.5,1ɑ.
      BSPAɑ:595.6,576.6ɑ.ɑ:3ɑ.ɑ:0.2,0.4ɑ.
```

```
*** PROGNOSEKORREKTUR ***
    ZUM ERGEBNIS:  0  5


         VARIABLE:        BSP

         URSPR.           KORR.
         PROGN.           PROGN.

7401     613.173          601.452
7501     600.397          591.828


         VARIABLE:        A I A

         URSPR.           KORR.
         PROGN.           PROGN.

7401     79.946           76.973
7501     81.143           78.906


         VARIABLE:        EF

         URSPR.           KORR.
         PROGN.           PROGN.

7401     186.947          191.304
7501     184.013          186.854


         VARIABLE:        PV

         URSPR.           KORR.
         PROGN.           PROGN.

7401     340.824          336.433
7501     350.347          346.856
STOP

ENDE STDHP  1.80
```

Footnotes

1 See [9].

2 See J. Cameron [4].

3 See for example the results of R. L. Cooper [6].

4 See also A. Mc Whorter [10].

5 See also C. W. J. Granger and P. Newbold [7].

6 See K. S. Åstroem [1, pp. 213].

7 See K. H. Raabe [13].

8 For further references see R. Hujer and R. Cremer [8, pp. 272].

[9] See K. S. Åstroem [1] or J. S. Meditch [11].

[10] See D. A. Belsley and E. Kuh [2].

[11] It should be noticed that the variables $y_1^n(t), \ldots, y_{N_p}^n(t)$ contain all those dependent variables that appear on the right hand side of the observation-model.

[12] It should be observed that we not have necessarily $N_a \leq N_p$. In case $N_a > N_p$ the $(N_a \times N_a)$-matrix $H\,V(t)'$ is not regular, so that the choice of $W(t)$ must ascertain that the matrix $(H\,V(t)\,H' + W(t))^{-1}$ used afterwards does exist.

[13] $[H(t)]_n$ is the n-th line of an $(N_a \times N_p)$-matrix $H(t)$ and $g_n(t)$ a mapping of $\mathbb{R}^K \times \mathbb{R}^{K'}$ in $\mathbb{R}^1$.

[14] For the matrix H can be assumed that $[H]_n \neq 0$ is true for all $n = 1, \ldots, N_a$. It can be shown that in the case of $[H]_n = 0$ for any n the corresponding n-th equation of the observation-model does not contribute to the correction, so that it can be dropped a priori.

[15] See to the possibilities of a smoothing at the problem of an estimation of time varying coefficients for example T. Cooley, B. Rosenberg and K. Wall [5].

[16] To demonstrate the procedure we refer to the possibilities of the MEBA-system. This EDP-system was developed by Institut für Angewandte Wirtschaftsforschung Tübingen within the 2nd and 3rd EDP-program promoted of the government of the FRG. See Bundesministerium für Wirtschaft [3]. With regard to this research, the basis-version of this EDP-system was extended by some experimental programs.

[17] To this model see Bundesministerium für Wirtschaft [3, pp. 21].

[18] For the various steps see Bundesministerium für Wirtschaft [3, especially pp. 53, pp. 67, pp. 111].

[19] The real observation values are:

	BSP	AIA	EF	PV
7401	595.6	69.4	190.9	337.8
7501	576.6	69.6	192.4	344.8

[20] We do not deal with the reproduction of a kind of a sensitivity-analysis in the form of a variation of the input-parameters for the specification of the reliability of information.

References

[1] Åstroem, K. S. (1970), *Introduction to stochastic control*, New York.

[2] Belsley, D. A. and Kuh, E. (1973), Time varying parameter structures: An overview, *Annals of Economic and Social Measurement* 2.

[3] Bundesministerium für Wirtschaft (1978), *Ökonometrische Methodenbank - Neues Instrument im Bereich gesamtwirtschaftlicher Analysen und Projektionen*, Bonn.

[4] Cameron, J. (1978), The economic modelers vie for Washington's ear, *Fortune* 20. Nov. 1978.

[5] Cooley, T., Rosenberg, B. and Wall, K. (1977), A note on an optimal smoothing for time varying coefficient problem, *Annals of Economic and Social Measurement* 6.

[6] Cooper, R. L. (1972), The predictive performance of quarterly econometric models of the United States, in: B. G. Hickman (ed.), *Econometric models of cyclical behavior*, New York.

[7] Granger, C. W. J. and Newbold, P. (1977), The time series approach to econo-
     metric model building, in: Federal Reserve Bank of Minneapolis, *New methods in
     business cycle research: Proceedings from a conference*, Minneapolis.
[8] Hujer, R. and Cremer, R. (1978), *Methoden der empirischen Wirtschaftsforschung*,
     München.
[9] Kommission der Europäischen Gemeinschaften (1971), *Kolloquium über die Metho-
     den zur Aufstellung des Wirtschaftsbudgets innerhalb der Gemeinschaft*, Brüssel.
[10] Mc Whorter, A., *Time series forecasting using the Kalman-filter: An empirical study*,
     mimeographed manuscript.
[11] Meditch, J. S. (1969), *Stochastic optimal linear estimation and control*, New York.
[12] Menges, G. (1978), Das Heidelberger Modell (und ein paar grundsätzliche Refle-
     xionen), in: J. Frohn (ed.), *Makroökonometrische Modelle für die Bundesrepublik
     Deutschland*, Göttingen.
[13] Raabe, K. H. (1969), *Prognose und Projektionen der kurzfristigen Wirtschaftsent-
     wicklung*, Bonn.

Time Series Analysis and Hypotheses Search Procedures

by

Hermann Garbers*

Zürich

Zusammenfassung:
Der Beitrag diskutiert Probleme einer empirischen aber nichtexperimentellen Wissenschaft, deren Daten in der Form von Zeitreihen vorliegen.

Summary:
The paper discusses some points about the search of hypotheses in the framework of a non-experimental science with time series data.

1. The data of an economist consist to a large part of time series. Empirical economic research is therefore to a large extent time series analysis.

2. Most economic time series are given in a non-experimental situation and are non repeatable. If these time series are modelled as realisations of stochastic processes, there exists only a single time series on which to base the investigation of the family of distribution functions. Therefore, in empirical economic research one discusses only the two first moment functions.

3. The empirical description of stochastic processes by moment functions requires of course
1. the existence and
2. the estimability of these functions.

The first assumption seems to be not a serious one, inspite of the fact that the existence of the integral

$$\int_{-\infty}^{+\infty} x^2 f(x)\,dx$$

* I would like to thank Prof. Wolfgang Stolper and members of the "Institut für Empirische Wirtschaftsforschung, Zürich" for helpful assistance.

with a density function $f(\cdot)$ is not selfevident and Fama's [1] finding of some evidence that at least the price formation processes on speculative markets have no finite variance.

The second assumption leads to the restricted class of covariance ergodic processes.

4. Weakly stationary stochastic processes with absolutely summable covariance functions are ergodic with a finite variance. Therefore, they are widely used to model economic time series. Very often, the original time series must be transformed in an attempt to make them stationary. In this context the choice of transformation is often done mechanically and after simply looking at the data. Quite often it consists only in forming logarithms and/or first differences.

5. The last sentence already implies that the economist's search for a model depends on the data. Contrary to classical statistics and experimental science we do not have a situation in which the model is known, but not its parameters. The construction of a model, the estimation of the parameters, and the checking of the hypothesis have to be done more or less simultaneously and quite often with the same set of data. We want to have a look at a widely used research strategy to see how this problem is treated practically.

6. Let us assume that our data are the partially known results of an "experiment" done by "nature". From our observations we try to deduce the nature of the experiment and determine, whether we can gain new insights from the results.

In analysing the experimental conditions that generated the given time series, one usually starts with some kind of economic theory. This theory represents to a substantial extent the prior knowledge of the researcher. It suggests which variables are of special interest. These suggestions usually contain uncertainties for instance, with respect to the time-lags of the explanatory variables of an equation. The discussion proceeds then very often by stepwise regressions: $Y(t)$ is regressed amongst others on $X(t)$, $X(t-1)$, ... till stochastically insignificant $X(t-k)$-variables occur. Apparently zero-restrictions of the parameters of $X(t-k)$, $X(t-k-1)$ are assumed at step k. The restrictions are then successively relaxed. This procedure obviously implies that restrictions are possible. Hence "let the data speak".

7. What is valid for the specification of the dynamic structure often holds good also for the choice of the variables of the equation: they are made to depend on the data by certain t- and F-values of classical statistics.

An impressive argument against this research strategy is made by Granger and Newbold [2]. In a simulation study they repeatedly generated realisations of e.g., five *independent* random walk processes and regressed one of them on the others with the OLS-procedure. "Highly significant" t- and R^2-values caused by the low frequency components in all processes resulted.

8. To avoid these spurious regression results, one could – at least at first sight – think of a detrending of the processes which precedes the regression. Yet, with respect to the relations between the variables, a type-one-error could be introduced in this manner in addition to the type-two-error of Granger and Newbold. It is partly for this reason that a hypotheses search strategy must be found in which such a preceding detrending is only undertaken *after* OLS-estimation has yielded a "too low" Durbin-Watson d-statistic[1].

9. In the research strategy just mentioned, a rejection of the null hypothesis "the error terms are not correlated" leads to a decision which is based on an *a priori* reasoning, just as it was decided before on the basis of t-statistics whether to include or exclude particular variables from certain relations.

It must felt to be particularly annoying that in this research strategy there seems to be no room for diagnosting the reasons for the rejection of a null hypothesis which might in turn lead to a change of the original hypothesis.

It can therefore be concluded:

In the research strategy developed above, information from data is combined with certain theoretical presuppositions of the researcher. The integration of these presuppositions is mainly done on an *ad hoc* basis. In this research strategy, there is no learning process of the researcher which accompanies the hypotheses search procedure.

10. The obvious *ad hoc* nature of combining *a priori* knowledge with observations is evidenced by the following practice. A relationship between variables, once found, is almost always justified *ex post* with additional evidence not contained in the data base. For example, an empirical study of the impact of sales on investments may fail to yield a geometrically declining lag pattern. This observation is then interpreted in terms of the two stage character of investment. Cross section data suggest that firms evaluate sales in one way during the planning stage, but differently during the realization phase of the investment.

Rationalizations of this sort are not without problems: "... a great wealth of prior information is available to aid in the interpretation of current observations. Unfortunately, prior information is not accessed costlessly ... The recalled information may ... be a version of the past that is distorted to suit the present purposes." [4; p. 308].

11. The discussion may have created the impression that a Bayesian research strategy is advocated in this paper. The Bayesian strategy in fact integrates the analyst and his *a priori* knowledge into the analysis in a transparent manner; moreover, it contains a theory of learning.

First, the problems with this strategy lie with the delineation of the set on which e. g. the so-called prior distribution is defined [7; p. 243]. That space should ideally contain the set of all models which are compatible with the outcome of a given experiment. However, that set cannot be constructed in real situations and may in fact not be well-defined at all[2]. For any such collection of models (hypotheses) can be easily enlarged by thinking of some other hypothesis that might hold true. The probability of the corresponding event must remain undefined. As Stegmüller [7; p. 244] puts it, "I therefore tend to conclude ... that the subjectivist theory (of probability) fails, because of the problems posed by the unexpected hypothesis"[3].

12. Clearly the set of hypotheses introduced in section 11 would have to be constructed prior to any analysis of the data. Otherwise, choice of the likelihood factor as well as the prior distribution may be influenced by the observations at hand. Conversely, if the search for hypotheses is partially guided by the data, this must be interpreted as an attempt to restrict the set of conceivable hypotheses to the class of "relevant" ones. Of course, data already exploited in such a search are only of limited value, when it comes to testing those same hypotheses[4]. This procedure very likely introduces a positive bias. Hence, an evaluation for example based on the t-statistic has little to do anymore with true t-test, and it would be better to speak of a diagnostic check rather than of a test.

Remark: The problem just alluded to, is also relevant to the issue of pretesting [6; chapter 4]. Unfortunately, the discussion of that issue contributes nothing to our problem, because it bears on the role of uncertain *a priori* knowledge within the framework of a given model.

Let

$$Y = X\beta + U$$

denote a classical normal regression model with a set of linear restrictions

(1) $R\beta = r$.

Here R is a fixed known (q, k) matrix of full column rank and r a fixed vector. Let there be uncertainty as to whether in fact β satisfies the restrictions (1). Then, β can be estimated either using e.g. the OLS estimator $\hat{\beta}$ or the modified OLS estimator $\hat{\beta}_R$ which takes (1) for granted. Now pretest estimators arise when the analyst chooses between $\hat{\beta}$ and $\hat{\beta}_R$ according to the outcome of a test (an F-test) of (1), but continues to use the same data for reestimation and subsequent statistical testing.

13. The reference to the "unexpected hypothesis" on the one hand and the data dependence of the search process on the other do point out that empirical economic investigation should be at least to some degree exploratory data analysis.

Certainly an economist who does not expect much additional information from time series *relative to his prior knowledge,* will be reserved in using such kind of data analysis. He will rather tend to use the data to confirm his knowledge. This will seem the more probable, the more hypotheses components are introduced only to make the model complete. A violation of these (from the point of view of the investigator) secondary hypotheses will merely motivate him to choose another (as he believes) more efficient estimator for the parameters of primary interest. In this context one may think of modelling some relationship using a classical regression model where the assumptions about the disturbance process are considered to be of secondary importance.

If the Durbin-Watson statistic were to turn out comparatively low the investigator, convinced of his prior knowledge, would merely formulate a new model using a generalized regression framework.

14. An investigator convinced of the correctness of his knowledge, will surely run into problems as soon as he meets other economists equally convinced of their different presupposition about the same subject. This argument was used by Hendry and Davidson [3] to explain the fact that even after decades of discussion there still exist different empirically supported theories about the relation between aggregate consumer spending and disposable income of households. Each of these theories can make potential use of the same data set. But each of them uses only a part of the data depending on the different presuppositions. Instead of an exploratory data analysis, some kind of analysis based on more or less arbitrary restrictions is made. Hendry and Davidson: "... nobody finds phenomena for which they never search".

15. An essential motivation for going through this arbitrary-looking way of data analysis described in section 14 may be the fear of a large set of noninformative data. How could this arbitrariness be confined without denying at the same time the existence of prior knowledge? An answer to this question might be the integration of different pieces of prior knowledge into an enlarged hypothesis. In this manner, Hendry and Davidson were able to represent four different approaches explaining the relationship between aggregate consumer spending (C) and disposable income (Y) of households as special cases of the following model:

$$C_t = \xi_0 + \xi_1 Y_t + \xi_2 Y_{t-4} + \xi_3 \Delta_1 Y_t + \xi_4 \Delta_1 Y_{t-4} + \xi_5 C_{t-4}$$

$$+ \sum_{j=1}^{3} \xi_{5+j} Q_{jt} + \xi_9 D_t^0 + \xi_{10} D_{t-4}^0 + \xi_{11} C_{t-1} + \xi_{12} t$$

$$+ \sum_{j=1}^{3} \xi_{12+j} Q_{jt} t + \varepsilon_t .$$

Notice: The $Q_{jt}(\cdot)$'s are seasonal "dummy" variables, t is a time variable, the D_t^0's are some special "dummy"-variables.

The four theories represent insofar special cases of this extensive model as they imply different restrictions on the parameters which one may try to confirm e.g. by using t-statistics.

16. Proceeding in the way of Hendry and Davidson restricts the arbitrariness of the *a priori* information and strengthens the significance of the data. However, this strategy too, is subject to criticism. Quite often, it might lead to multi collinear relations and consequently, among other things, to potentially ill-conditioned systems of normal equations and numerically insignificant regression results.

17. Proceeding according to Hendry and Davidson also raises the question, if and at what cost different approaches can be represented as special cases of the same comprehensive theory. Considering the fact that some of these theories deal with levels of variables, others with their differences or growth rates and still others with their logarithms, the problem seems to be rather difficult.

18. The difficulties mentioned so far rather suggest that an explorative data analysis approach should be used in empirical economic research. Corresponding to section 1 to 4 of this paper and to the well known decomposition theorem of H. Wold, this kind of research strategy deals mainly with multidimensional ARIMA-processes of the form

$$A(L)(1-L)^d(1-L^S)^D Y(t) = B(L)\varepsilon(t)^{\ 5}.$$

19. Due to the almost complete neglect of any *a priori* information, such vectorial ARIMA-processes are not particularly suitable to model economic relations. One-dimensional ARIMA-models have, nevertheless, led to excellent short- and medium-term forecasts. The integration of economic *a priori* information may, therefore, be realized in an additive way: In the identification phase of a model, the residuals

$$A(L)(1-L)^{d_1}(1-L^S)^{D_1} Y(t)$$

could, for instance, be successively regressed on some *a priori* selected variables $X(t)$, where the $X(t)$-processes, however, were previously made stationary by the transformation

$$(1-L)^{d_2}(1-L^S)^{D_2}\ .$$

20. Once again, the mechanical transformations of non-stationary processes have to be considered. In the context of ARIMAX-processes which were introduced in section 19, these transformations can be replaced by methods which are instead based on considerations of economic theory. This is due to the fact that the non-stationarity of the Y-variable is caused by the non-stationarity of the X-process. In the case of a single X-variable, Nerlove et al. [5; chapter XI] recommend a strategy, as introduced in section 19, to make the two processes stationary. The common terms of the two transformations are then eliminated before the resulting variables are modelled by an ARIMAX-process.

21. ARIMAX-models gain a certain attractivity indeed which is not only due to the fact that data-analysis is exploratory work, but also because the analyst has already some knowledge at his disposal. This may be explained also by Zellner and Palm's [8] argument:

If the vector of economic variables $Y(t)$ (if necessary transformed) in

$$(*) \qquad B(L) Y(t) = A(L) \varepsilon(t)$$

may be decomposed e.g. for reasons of an economic theory into a subvector $Z(\cdot)$ of endogenous and a subvector $X(\cdot)$ of exogenous variables, then it follows with an appropriate decomposition of the matrices that:

$$\begin{bmatrix} B_{11}(L) & B_{12}(L) \\ B_{21}(L) & B_{22}(L) \end{bmatrix} \begin{bmatrix} Z(\cdot) \\ X(\cdot) \end{bmatrix}$$
$$= \begin{bmatrix} A_{11}(L) & A_{12}(L) \\ A_{21}(L) & A_{22}(L) \end{bmatrix} \begin{bmatrix} \varepsilon_1(\cdot) \\ \varepsilon_2(\cdot) \end{bmatrix}.$$

Since $X(\cdot)$ is exogenous

$$B_{21}(L) = 0 \quad \text{and} \quad A_{12}(L) = A_{21}(L) = 0,$$

in order to eliminate any influence of $Z(\cdot)$ on to $X(\cdot)$, even indirect effect over $\varepsilon_1(\cdot)$. The relation $(*)$ may then be written as:

$$B_{11}(L) Z(\cdot) + B_{12}(L) X(\cdot) = A_{11}(L) \varepsilon_1(\cdot)$$
$$B_{22}(L) X(\cdot) = A_{22}(L) \varepsilon_2(\cdot).$$

The second subsystem of equations is often neglected by economists for whom they seem to be economically uninteresting. Therefore the analysis usually concentrates on the so-called *structural equations*:

$$B_{11}(L) Z(t) + B_{12}(L) X(t) = A_{11}(L) \varepsilon_1(t).$$

Given the existence of the inverse of $B_{11}(L)$

$$Z(t) = -B_{11}^{-1}(L) B_{12}(L) X(t) + B_{11}^{-1}(L) A_{11}(L) \varepsilon_1(t)$$

and that implies

$$|B_{11}(L)| Z(t) = -B_{11}^{*}(L) B_{12}(L) X(t)$$
$$+ B_{11}^{*}(L) A_{11}(L) \varepsilon_1(t)$$

with the adjoint matrix $B_{11}^{*}(L)$ of $B_{11}(L)$. But with these transformations we end up with a system of (cross-correlated) ARMAX-processes.

22. The aim of the previous reasoning was to integrate already existing knowledge into ARMAX-models. However, we should remember that a modelbuilder can freely choose what set of questions his model is designed to answer. Hence ARMAX-models may be more or less simple[6].

23. Concerning the $X(\cdot)$ variables we have mentioned their additivity. Looking at the specification of these variables we - unlike Hendry and Davidson - rather think of an "intersection" than of a "unification" of theories. But the problem of ill-conditioned normal equations is not eliminated in either case. It may well be that with $|B_{11}(L)|$ and $B_{11}^*(L)$, we have polynomials in L of a high degree. If $Z(\cdot)$ has, say, ten components each containing an autoregressive subprocess of at least second order, then the order of $|B_{11}(L)|$ itself is even greater than 20.

24. Within the macro-economic framework the arguments favour highly aggregated models. But, due to the non-homogeneous variables in these models, may it not then happen that there is neither an acceptable theory nor a set of conditions for its existence? That is the new question to answer.

Footnotes

[1] Granger's and Newbold's exposition also contains a similar suggestion.

[2] Presumably, the costs involved in this construction could not be neglected either.

[3] Translation by the author.

[4] As a practical example, consider the following situation. A model consisting of simultaneous relations between interdependent variables is being constructed. In each case, OLS estimates are used to find a preferred hypothesis. Finally, estimation of the entire system proceeds with the same data and is accompanied by the "right" statistical tests - without any reference to the severely reduced power of these tests.

[5] Using the common notation.

[6] The analogy to the discussion about the reduced form of traditional econometric models cannot be overlooked.

References

[1] Fama, E. F. (1965), The behavior of stock market prices, *Journal of Business* 38, 34–105.
[2] Granger, C. W. J. and Newbold, P. (1974), Spurious regressions in econometrics, *Journal of Econometrics* 2, 111–120.
[3] Hendry, D. F. and Davidson, J. E. H. (1977), *Econometric modelling of the aggregate time series relationship between consumers' expenditure and income in the United King-*

dom, Paper presented at the annual meeting of the Econometric Society, Vienna: 1977.

[4] Leamer, E. (1978), *Specification Searches*, New York: Wiley.
[5] Nerlove, M., Grether, D. M. and Carvalho, J. L. (1979), *Analysis of economic time series*, New York: Academic Press.
[6] Seiterle, H. (1979), *Probleme der empirischen Wirtschaftsforschung*, Zürich.
[7] Stegmüller, W. (1973), *Probleme und Resultate der Wissenschaftstheorie und Analytischen Philosophie*, Band IV, Studienausgabe Teil D, Berlin: Springer Verlag.
[8] Zellner, A. and Palm, F. (1974), Time series analysis and simultaneous equation econometric models, *Journal of Econometrics* 2, 17–54.